Advanced Calculus
Demystified

Demystified Series

Accounting Demystified
Advanced Statistics Demystified
Algebra Demystified
Alternative Energy Demystified
Anatomy Demystified
ASP.NET 2.0 Demystified
Astronomy Demystified
Audio Demystified
Biology Demystified
Biotechnology Demystified
Business Calculus Demystified
Business Math Demystified
Business Statistics Demystified
C++ Demystified
Calculus Demystified
Chemistry Demystified
College Algebra Demystified
Corporate Finance Demystified
Data Structures Demystified
Databases Demystified
Differential Equations Demystified
Digital Electronics Demystified
Earth Science Demystified
Electricity Demystified
Electronics Demystified
Environmental Science Demystified
Everyday Math Demystified
Forensics Demystified
Genetics Demystified
Geometry Demystified
Home Networking Demystified
Investing Demystified
Java Demystified
JavaScript Demystified
Linear Algebra Demystified
Macroeconomics Demystified
Management Accounting Demystified

Math Proofs Demystified
Math Word Problems Demystified
Medical Billing and Coding Demystified
Medical Terminology Demystified
Meteorology Demystified
Microbiology Demystified
Microeconomics Demystified
Nanotechnology Demystified
Nurse Management Demystified
OOP Demystified
Options Demystified
Organic Chemistry Demystified
Personal Computing Demystified
Pharmacology Demystified
Physics Demystified
Physiology Demystified
Pre-Algebra Demystified
Precalculus Demystified
Probability Demystified
Project Management Demystified
Psychology Demystified
Quality Management Demystified
Quantum Mechanics Demystified
Relativity Demystified
Robotics Demystified
Signals and Systems Demystified
Six Sigma Demystified
SQL Demystified
Statics and Dynamics Demystified
Statistics Demystified
Technical Math Demystified
Trigonometry Demystified
UML Demystified
Visual Basic 2005 Demystified
Visual C# 2005 Demystified
XML Demystified

Advanced Calculus
Demystified

David Bachman

New York   Chicago   San Francisco   Lisbon   London
Madrid   Mexico City   Milan   New Delhi   San Juan
Seoul   Singapore   Sydney   Toronto

The McGraw·Hill Companies

Library of Congress Cataloging-in-Publication Data

Bachman, David.
   Advanced calculus demystified / David Bachman.
      p. cm. – (Demystified series)
   Includes index.
   ISBN-13: 978-0-07-148121-2 (alk. paper)
   ISBN-10: 0-07-148121-4 (alk. paper)
   1. Calculus. I. Title.
   QA303.2.B23 2007
   515–dc22
                                    2007016733

McGraw-Hill books are available at special quantity discounts to use as premiums and sales promotions, or for use in corporate training programs. For more information, please write to the Director of Special Sales, Professional Publishing, McGraw-Hill, Two Penn Plaza, New York, NY 10121-2298. Or contact your local bookstore.

Advanced Calculus Demystified

234567890 DOC DOC 01987

ISBN-13:   978-0-07-148121-2
ISBN-10:   0-07-148121-4

Sponsoring Editor Judy Bass	**Indexer** Valerie Perry
Editorial Supervisor Janet Walden	**Production Supervisor** Jean Bodeaux
Project Manager Gita Raman	**Composition** International Typesetting and Composition
Technical Editor Steven Krantz	**Illustration** International Typesetting and Composition
Copy Editor Surendra Nath Shivam	**Art Director, Cover** Jeff Weeks
Proofreader Yumnam Ojen	**Cover Illustration** Lance Lekander

To Stacy

ABOUT THE AUTHOR

David Bachman, Ph.D. is an Assistant Professor of Mathematics at Pitzer College, in Claremont, California. His Ph.D. is from the University of Texas at Austin, and he has taught at Portland State University, The University of Illinois at Chicago, as well as California Polytechnic State University at San Luis Obispo. Dr. Bachman has authored one other textbook, as well as 11 research papers in low-dimensional topology that have appeared in top peer-reviewed journals.

CONTENTS

Contents

PREFACE

In the first year of calculus we study limits, derivatives, and integrals of functions with a single input, and a single output. The transition to *advanced calculus* is made when we generalize the notion of "function" to something which may have multiple inputs and multiple outputs. In this more general context limits, derivatives, and integrals take on new meanings and have new geometric interpretations. For example, in first-year calculus the derivative represents the slope of a tangent line at a specified point. When dealing with functions of multiple variables there may be many tangent lines at a point, so there will be many possible ways to differentiate.

The emphasis of this book is on developing enough familiarity with the material to solve difficult problems. Rigorous proofs are kept to a minimum. I have included numerous detailed examples so that you may see how the concepts really work. All exercises have detailed solutions that you can find at the end of the book. I regard these exercises, along with their solutions, to be an integral part of the material.

The present work is suitable for use as a stand-alone text, or as a companion to any standard book on the topic. This material is usually covered as part of a standard calculus sequence, coming just after the first full year. Names of college classes that cover this material vary greatly. Possibilities include *advanced calculus*, *multivariable calculus*, and *vector calculus*. At schools with semesters the class may be called *Calculus III*. At quarter schools it may be *Calculus IV*.

The best way to use this book is to read the material in each section and then try the exercises. If there is any exercise you don't get, make sure you study the solution carefully. At the end of each chapter you will find a quiz to test your understanding. These short quizzes are written to be similar to one that you may encounter in a classroom, and are intended to take 20–30 minutes. They are not meant to test every idea presented in the chapter. The best way to use them is to study the chapter until you feel confident that you can handle anything that may be asked, and then try the quiz. You should have a good idea of how you did on it after looking at the answers. At the end of the text there is a final exam similar to one which you would find at the conclusion of a college class. It should take about two hours to complete. Use it as you do the quizzes. Study all of the material in the book until you feel confident, and then try it.

Advanced calculus is an exciting subject that opens up a world of mathematics. It is the gateway to linear algebra and differential equations, as well as more advanced mathematical subjects like analysis, differential geometry, and topology. It is essential for an understanding of physics, lying at the heart of electro-magnetics, fluid flow, and relativity. It is constantly finding new use in other fields of science and engineering. I hope that the exciting nature of this material is conveyed here.

ACKNOWLEDGMENTS

The author thanks the technical editor, Steven G. Krantz, for his helpful comments.

CHAPTER 1

Functions of Multiple Variables

1.1 Functions

The most common mental model of a *function* is a machine. When you put some input in to the machine, you will always get the same output. Most of first year calculus dealt with functions where the input was a single real number and the output was a single real number. The study of advanced calculus begins by modifying this idea. For example, suppose your "function machine" took *two* real numbers as its input, and returned a single real output? We illustrate this idea with an example.

EXAMPLE 1-1
Consider the function

$$f(x, y) = x^2 + y^2$$

For each value of x and y there is one value of $f(x, y)$. For example, if $x = 2$ and $y = 3$ then

$$f(2, 3) = 2^2 + 3^2 = 13$$

One can construct a table of input and output values for $f(x, y)$ as follows:

x	y	$f(x, y)$
0	0	0
1	0	1
0	1	1
1	1	2
1	2	5
2	1	5

Problem 1 *Evaluate the function at the indicated point.*

1. $f(x, y) = x^2 + y^3$; $(x, y) = (3, 2)$
2. $g(x, y) = \sin x + \cos y$; $(x, y) = (0, \frac{\pi}{2})$
3. $h(x, y) = x^2 \sin y$; $(x, y) = (2, \frac{\pi}{2})$

Unfortunately, plugging in random points does not give much enlightenment as to the behavior of a function. Perhaps a more visual model would help....

1.2 Three Dimensions

In the previous section we saw that plugging random points in to a function of two variables gave almost no enlightening information about the function itself. A far superior way to get a handle on a particular function is to picture its *graph*. We'll get to this in the next section. First, we have to say a few words about where such a graph exists.

Recall the steps required to graph a function of a single variable, like $g(x) = 3x$. First, you set the function equal to a new variable, y. Then you plot all the points (x, y) where the equation $y = g(x)$ is true. So, for example, you would not plot $(0, 2)$ because $0 \neq 3 \cdot 2$. But you would plot $(2, 6)$ because $6 = 3 \cdot 2$.

The same steps are required to plot a function of two variables, like $f(x, y)$. First, you set the function equal to a new variable, z. Then you plot all of the points (x, y, z) where the function $z = f(x, y)$ is true. So we are forced to discuss what it means to plot a point with three coordinates, like (x, y, z).

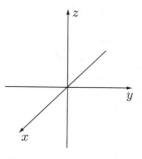

Figure 1-1   Three mutually perpendicular axes, drawn in perspective

    Coordinate systems will play a crucial role in this book, so although most readers will have seen this, it is worth spending some time here. To plot a point with two coordinates such as $(x, y) = (2, 3)$ the first step is to draw two perpendicular axes and label them x and y. Then locate a point 2 units from the origin on the x-axis and draw a vertical line. Next, locate a point 3 units from the origin on the y-axis and draw a horizontal line. Finally, the point $(2, 3)$ is at the intersection of the two lines you have drawn.

    To plot a point with three coordinates the steps are just a bit more complicated. Let's plot the point $(x, y, z) = (2, 3, 2)$. First, draw *three* mutually perpendicular axes. You will immediately notice that this is impossible to do on a sheet of paper. The best you can do is two perpendicular axes, and a third at some angle to the other two (see Figure 1-1). With practice you will start to see this third axis as a perspective rendition of a line coming out of the page. When viewed this way it will seem like it is perpendicular.

    Notice the way in which we labeled the axes in Figure 1-1. This is a convention, i.e., something that mathematicians have just agreed to always do. The way to remember it is by the *right hand rule*. What you want is to be able to position your right hand so that your thumb is pointing along the z-axis and your other fingers sweep from the x-axis to the y-axis when you make a fist. If the axes are labeled consistent with this then we say you are using a *right handed coordinate system.*

    OK, let's now plot the point $(2, 3, 2)$. First, locate a point 2 units from the origin on the x-axis. Now picture a *plane* which goes through this point, and is perpendicular to the x-axis. Repeat this for a point 3 units from the origin on the y-axis, and a point 2 units from the origin on the z-axis. Finally, the point $(2, 3, 2)$ is at the intersection of the three planes you are picturing.

    Given the point (x, y, z) one can "see" the quantities x, y, and z as in Figure 1-2. The quantity z, for example, is the distance from the point to the xy-plane.

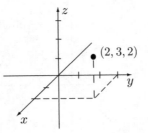

Figure 1-2   Plotting the point (2, 3, 2)

Problem 2 *Which of the following coordinate systems are right handed?*

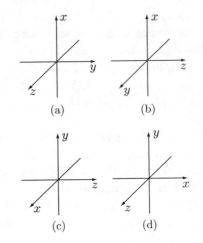

Problem 3 *Plot the following points on one set of axes:*

1. (1, 1, 1)
2. (1, −1, 1)
3. (−1, 1, −1)

1.3 Introduction to Graphing

We now turn back to the problem of visualizing a function of multiple variables. To graph the function $f(x, y)$ we set it equal to z and plot all of the points where the equation $z = f(x, y)$ is true. Let's start with an easy example.

EXAMPLE 1-2

Suppose $f(x, y) = 0$. That is, $f(x, y)$ is the function that always returns the number 0, no matter what values of x and y are fed to it. The graph of $z = f(x, y) = 0$ is then the set of all points (x, y, z) where $z = 0$. This is just the xy-plane.

Similarly, now consider the function $g(x, y) = 2$. The graph is the set of all points where $z = g(x, y) = 2$. This is a plane parallel to the xy-plane at height 2.

We first learn to graph functions of a single variable by plotting individual points, and then playing "connect-the-dots." Unfortunately this method doesn't work so well in three dimensions (especially when you are trying to depict three dimensions on a piece of paper). A better strategy is to slice up the graph by various planes. This gives you several curves that you can plot. The final graph is then obtained by assembling these curves.

The easiest slices to see are given by each of the coordinate planes. We illustrate this in the next example.

EXAMPLE 1-3

Let's look at the function $f(x, y) = x + 2y$. To graph it we must decide which points (x, y, z) make the equation $z = x + 2y$ true. The xz-plane is the set of all points where $y = 0$. So to see the intersection of the graph of $f(x, y)$ and the xz-plane we just set $y = 0$ in the equation $z = x + 2y$. This gives the equation $z = x$, which is a line of slope 1, passing through the origin.

Similarly, to see the intersection with the yz-plane we just set $x = 0$. This gives us the equation $z = 2y$, which is a line of slope 2, passing through the origin.

Finally, we get the intersection with the xy-plane. We must set $z = 0$, which gives us the equation $0 = x + 2y$. This can be rewritten as $y = -\frac{1}{2}x$. We conclude this is a line with slope $-\frac{1}{2}$.

The final challenge is to put all of this information together on one set of axes. See Figure 1-3. We see three lines, in each of the three coordinate planes. The graph of $f(x, y)$ is then some shape that meets each coordinate plane in the required line. Your first guess for the shape is probably a plane. This turns out to be correct. We'll see more evidence for it in the next section.

Problem 4 *Sketch the intersections of the graphs of the following functions with each of the coordinate planes.*

   1. $2x + 3y$
   2. $x^2 + y$
   3. $x^2 + y^2$

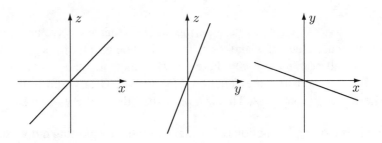

Figure 1-3 The intersection of the graph of $x + 2y$ with each coordinate plane is a line through the origin

4. $2x^2 + y^2$
5. $\sqrt{x^2 + y^2}$
6. $x^2 - y^2$

1.4 Graphing Level Curves

It's fairly easy to plot the intersection of a graph with each coordinate plane, but this still doesn't always give a very good idea of its shape. The next easiest thing to do is sketch some level curves. These are nothing more than the intersection of the graph with horizontal planes at various heights. We often sketch a "bird's eye view" of these curves to get an initial feeling for the shape of a graph.

EXAMPLE 1-4
Suppose $f(x, y) = x^2 + y^2$. To get the intersection of the graph with a plane at height 4, say, we just have to figure out which points in $\mathbb{R}^3$ satisfy $z = x^2 + y^2$ and $z = 4$. Combining these equations gives $4 = x^2 + y^2$, which we recognize as the equation of a circle of radius 2. We can now sketch a view of this intersection from above, and it will look like a circle in the xy-plane. See Figure 1-4.

The reason why we often draw level curves in the xy-plane as if we were looking down from above is that it is easier when there are many of them. We sketch several such curves for $z = x^2 + y^2$ in Figure 1-5.

You have no doubt seen level curves before, although they are rarely as simple as in Figure 1-5. For example, in Figure 1-6 we see a topographic map. The lines indicate constant elevation. In other words, these lines are the level curves for the function which gives elevation. In Figure 1-7 we have shown a weather map, with level curves indicating lines of constant temperature. You may see similar maps in a good weather report where level curves represent lines of constant pressure.

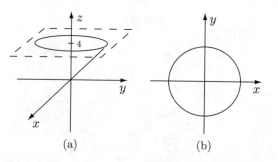

Figure 1-4   (a) The intersection of $z = x^2 + y^2$ with a plane at height 4. (b) A top view of the intersection

EXAMPLE 1-5

We now let $f(x, y) = xy$. The intersection with the xz-plane is found by setting $y = 0$, giving us the function $z = 0$. This just means the graph will include the x-axis. Similarly, setting $x = 0$ gives us $z = 0$ as well, so the graph will include the y-axis. Things get more interesting when we plot the level curves. Let's set $z = n$, where n is an integer. Solving for y then gives us $y = \frac{n}{x}$. This is a hyperbola in the first and third quadrant for $n > 0$, and a hyperbola in the second and fourth quadrant when $n < 0$. We sketch this in Figure 1-8.

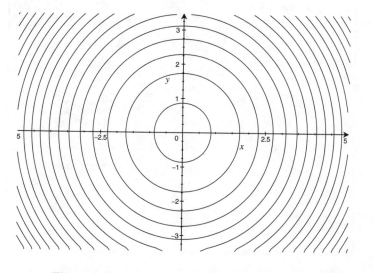

Figure 1-5   Several level curves of $z = x^2 + y^2$

Figure 1-6   A topographic map

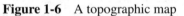

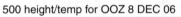

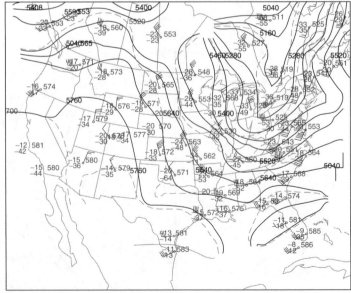

Figure 1-7   A weather map shows level curves

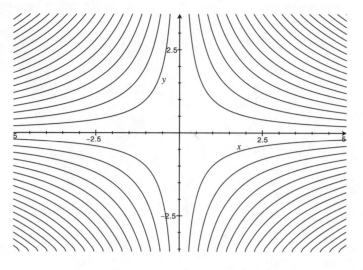

Figure 1-8   Level curves of $z = xy$

Problem 5 *Sketch several level curves for the following functions.*

1. $2x + 3y$
2. $x^2 + y$
3. $\sqrt{x^2 + y^2}$
4. $x^2 - y^2$

Problem 6 *The level curves for the following functions are all circles. Describe the difference between how the circles are arranged.*

1. $x^2 + y^2$
2. $\sqrt{x^2 + y^2}$
3. $\frac{1}{x^2+y^2}$
4. $\sin(x^2 + y^2)$

1.5 Putting It All Together

We have now amassed enough tools to get a good feeling for what the graphs of various functions look like. Putting it all together can be quite a challenge. We illustrate this with an example.

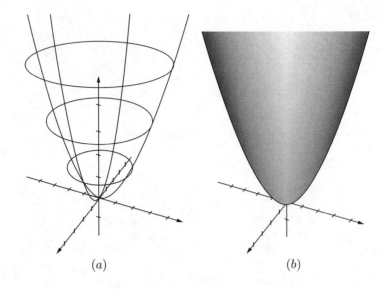

(a) (b)

Figure 1-9   Sketching the paraboloid $z = x^2 + y^2$

EXAMPLE 1-6

Let $f(x, y) = x^2 + y^2$. In Problem 4 you found that the intersections with the xz- and yz-coordinate planes were parabolas. In Example 1-4 we saw that the level curves were circles. We put all of this information together in Figure 1-9(a). Figure 1-9(b) depicts the entire surface which is the graph. This figure is called a *paraboloid*.

Graph sketching is complicated enough that a second example may be in order.

EXAMPLE 1-7

In Figure 1-10 we put together the level curves of $f(x, y) = xy$, found in Example 1-5, to form its graph. The three-dimensional shape formed is called a *saddle*.

Problem 7 *Use your answers to Problems 4 and 5 to sketch the graphs of the following functions:*

1. $2x + 3y$
2. $x^2 + y$
3. $2x^2 + y^2$
4. $\sqrt{x^2 + y^2}$
5. $x^2 - y^2$

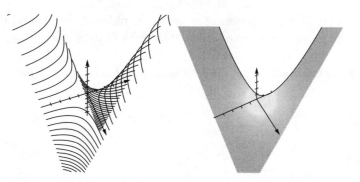

Figure 1-10   Several level curves of $z = xy$ piece together to form a saddle

1.6 Functions of Three Variables

There is no reason to stop at functions with two inputs and one output. We can also consider functions with three inputs and one output.

EXAMPLE 1-8
Suppose

$$f(x, y, z) = x + xy + yz^2$$

Then $f(1, 1, 1) = 3$ and $f(0, 1, 2) = 4$.

To graph such a function we would need to set it equal to some fourth variable, say w, and draw a picture in a space where there are four perpendicular axes, x, y, z, and w. No one can visualize such a space, so we will just have to give up on graphing such functions. But all hope is not lost. We can still describe surfaces in three dimensions that are the level sets of such functions. This is not quite as good as having a graph, but it still helps give one a feel for the behavior of the function.

EXAMPLE 1-9
Suppose

$$f(x, y, z) = x^2 + y^2 + z^2$$

To plot level sets we set $f(x, y, z)$ equal to various integers and sketch the surface described by the resulting equation. For example, when $f(x, y, z) = 1$ we have

$$1 = x^2 + y^2 + z^2$$

This is precisely the equation of a sphere of radius 1. In general the level set corresponding to $f(x, y, z) = n$ will be a sphere of radius $\sqrt{n}$.

Problem 8 *Sketch the level set corresponding to $f(x, y, z) = 1$ for the following functions.*

1. $f(x, y, z) = x^2 + y^2 - z^2$
2. $f(x, y, z) = x^2 - y^2 - z^2$

1.7 Parameterized Curves

In the previous sections of this chapter we studied functions which had multiple inputs, but one output. Here we examine the opposite scenario: functions with one input and multiple outputs. The input variable is referred to as the *parameter*, and is best thought of as time. For this reason we often use the variable t, so that in general such a function might look like

$$c(t) = (f(t), g(t))$$

If we fix a value of t and plot the two outputs we get a point in the plane. As t varies this point moves, tracing out a curve, C. We would then say C is a curve that is *parameterized by $c(t)$*.

EXAMPLE 1-10
Suppose $c(t) = (\cos t, \sin t)$. Then $c(0) = (1, 0)$ and $c\left(\frac{\pi}{2}\right) = (0, 1)$. If we continue to plot points we see that $c(t)$ traces out a circle of radius 1. Indeed, since

$$\cos^2 t + \sin^2 t = 1$$

the coordinates of $c(t)$ satisfy $x^2 + y^2 = 1$, the equation of a circle of radius 1. In Figure 1-11 we plot the circle traced out by $c(t)$, along with additional information which tells us what value of t yields selected point of the curve.

EXAMPLE 1-11
The function $c(t) = (\cos t^2, \sin t^2)$ also parameterizes a circle or radius 1, like the parameterization given in Example 1-10. The difference between the two parameterizations can be seen by comparing the spacing of the marked points in Figure 1-11 with those of Figure 1-12. If we think of t as time, then the parameterization

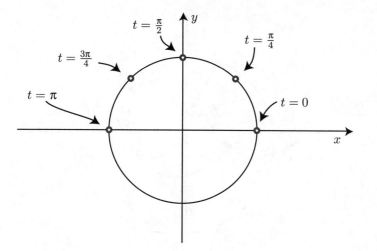

Figure 1-11   The function $c(t) = (\cos t, \sin t)$ parameterizes a circle of radius 1

depicted in Figure 1-12 represents a point moving around the circle faster and faster.

EXAMPLE 1-12

Now let $c(t) = (t \cos t, t \sin t)$. Plotting several points shows that $c(t)$ parameterizes a curve that spirals out from the origin, as in Figure 1-13.

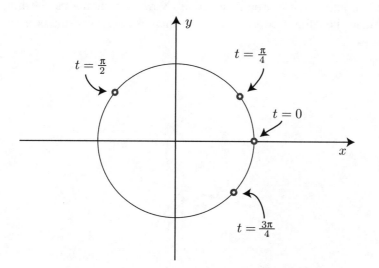

Figure 1-12   The function $c(t) = (\cos t^2, \sin t^2)$ parameterizes a circle of radius 1 in a different way

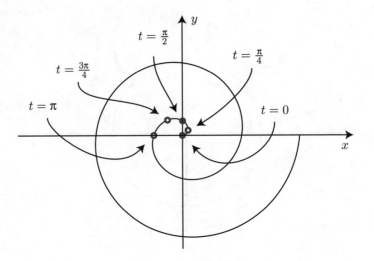

Figure 1-13 The function $c(t) = (t \cos t, t \sin t)$ parameterizes a spiral

Parameterizations can also describe curves in three-dimensional space, as in the next example.

EXAMPLE 1-13
Let $c(t) = (\cos t, \sin t, t)$. If the third coordinate were not there then this would describe a point moving around a circle. Now as t increases the height off of the xy-plane, i.e., the z-coordinate, also increases. The result is a spiral, as in Figure 1-14.

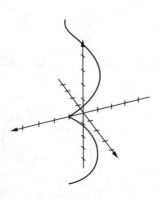

Figure 1-14 The function $c(t) = (\cos t, \sin t, t)$ parameterizes a curve that spirals around the z-axis

Problem 9 *Sketch the curves parameterized by the following:*

1. (t, t)
2. (t, t^2)
3. (t^2, t)
4. (t^2, t^3)
5. $(\cos 2t, \sin 3t)$

Problem 10 *The functions given in Examples 1-10 and 1-11 parameterize the same circle in different ways. Describe the difference between the two parameterizations for negative values of t.*

Problem 11 *Find a parameterization for the graph of the function $y = f(x)$.*

Problem 12 *Describe the difference between the following parameterized curves:*

1. $c(t) = (\cos t, \sin t, t^2)$
2. $c(t) = (\cos t, \sin t, \frac{1}{t})$
3. $c(t) = (t \cos t, t \sin t, t)$

Quiz

Problem 13

1. *Determine if the coordinate system pictured is left or right handed.*

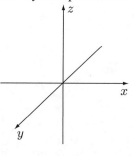

2. *Let $f(x, y) = \frac{y}{x^2+1}$.*

   a. *Sketch the intersections of the graph of $f(x, y)$ with the xy-plane, the xz-plane, and the yz-plane.*

   b. *Sketch the level curves for $f(x, y)$.*

   c. *Sketch the graph of $f(x, y)$.*

3. *Sketch the curve parameterized by $c(t) = (2 \cos t, 3 \sin t)$.*

CHAPTER 2

Fundamentals of Advanced Calculus

2.1 Limits of Functions of Multiple Variables

The study of calculus begins in earnest with the concept of a limit. Without this one cannot define derivatives or integrals. Here we undertake the study of limits of functions of multiple variables.

Recall that we say $\lim_{x \to a} f(x) = L$ if you can make $f(x)$ stay as close to L as you like by restricting x to be *close enough* to a. Just how close "close enough" is depends on how close you want $f(x)$ to be to L.

Intuitively, if $\lim_{x \to a} f(x) = L$ we think of the values of $f(x)$ as getting closer and closer to L as the value of x gets closer and closer to a. A key point is that it should not matter *how* the values of x are approaching a. For example, the function

$$f(x) = \frac{x}{|x|}$$

does not have a limit as $x \to 0$. This is because as x approaches 0 from the right the values of the function $f(x)$ approach 1, while the values of $f(x)$ approach -1 as x approaches 0 from the left.

The definition of limit for functions of multiple variables is very similar. We say

$$\lim_{(x,y)\to(a,b)} f(x, y) = L$$

if you can make $f(x, y)$ stay as close to L as you like by restricting (x, y) to be *close enough* to (a, b). Again, just how close "close enough" is depends on how close you want $f(x, y)$ to be to L.

Once again, the most useful way to think about this definition is to think of the values of $f(x, y)$ as getting closer and closer to L as the point (x, y) gets closer and closer to the point (a, b). The difficulty is that there are now an infinite number of directions by which one can approach (a, b).

EXAMPLE 2-1
Suppose $f(x, y)$ is given by

$$f(x, y) = \frac{x}{x + y}$$

We consider $\lim_{(x,y)\to(0,0)} f(x, y)$.

First, let's see what happens as (x, y) approaches $(0, 0)$ along the x-axis. For all such points we know $y = 0$, and so

$$f(x, y) = \frac{x}{x + y} = \frac{x}{x} = 1$$

Now consider what happens as (x, y) approaches $(0, 0)$ along the y-axis. For all such points we have $x = 0$, and so

$$f(x, y) = \frac{x}{x + y} = \frac{0}{y} = 0$$

We conclude the values of $f(x, y)$ approach different numbers if we let (x, y) approach $(0, 0)$ from different directions. Thus we say $\lim_{(x,y)\to(0,0)} f(x, y)$ does not exist.

Showing that a limit does not exist can be very difficult. Just because you can find multiple ways to come at (a, b) so that the values of $f(x, y)$ approach the same number L does not necessarily mean $\lim\limits_{(x,y)\to(a,b)} f(x, y) = L$. There might be some way to approach (a, b) that you haven't tried that gives a different number. This is the key to the definition of limit. We say the function has a limit only when the values of $f(x, y)$ approach the same number no matter how (x, y) approaches (a, b). We illustrate this in the next two examples.

EXAMPLE 2-2

Suppose $f(x, y)$ is given by

$$f(x, y) = \frac{xy}{x^2 + y^2}$$

As we let (x, y) approach $(0, 0)$ along the x-axis (where $y = 0$) we have

$$f(x, y) = \frac{xy}{x^2 + y^2} = \frac{0}{x^2} = 0$$

Similarly, as we let (x, y) approach $(0, 0)$ along the y-axis (where $x = 0$) we have

$$f(x, y) = \frac{xy}{x^2 + y^2} = \frac{0}{y^2} = 0$$

But if we let (x, y) approach $(0, 0)$ along the line $y = x$ we have

$$f(x, y) = \frac{xy}{x^2 + y^2} = \frac{x^2}{2x^2} = \frac{1}{2}$$

So once again we find $\lim\limits_{(x,y)\to(0,0)} f(x, y)$ does not exist.

Our third example is the trickiest.

EXAMPLE 2-3

Let

$$f(x, y) = \frac{x^2 y}{x^4 + y^2}$$

As (x, y) approaches $(0, 0)$ along the x-axis (where $y = 0$) we have

$$f(x, y) = \frac{x^2 y}{x^4 + y^2} = \frac{0}{x^4} = 0$$

As (x, y) approaches $(0, 0)$ along the y-axis (where $x = 0$) we have

$$f(x, y) = \frac{x^2 y}{x^4 + y^2} = \frac{0}{y^2} = 0$$

If we approach $(0, 0)$ along the line $y = x$ we get

$$f(x, y) = \frac{x^2 y}{x^4 + y^2} = \frac{x^3}{x^4 + x^2} = \frac{x}{x^2 + 1}$$

As x approaches 0 we have

$$\lim_{x \to 0} \frac{x}{x^2 + 1} = 0$$

So far it is looking like perhaps

$$\lim_{(x,y) \to (0,0)} \frac{x^2 y}{x^4 + y^2} = 0$$

since, as (x, y) approaches $(0, 0)$ along the x-axis, the y-axis, and the line $y = x$, the values of $f(x, y)$ approach 0. But what happens if we let (x, y) approach $(0, 0)$ along the *curve* $y = x^2$? In this case

$$f(x, y) = \frac{x^2 y}{x^4 + y^2} = \frac{x^4}{x^4 + x^4} = \frac{1}{2}$$

We can evaluate the limit of this function as x approaches 0 by dividing the numerator and denominator by x^4.

$$\lim_{x \to 0} \frac{x^4}{x^2 + x^4} = \lim_{x \to 0} \frac{1}{\frac{1}{x^2} + 1} = 1$$

So again the limit does not exist.

Problem 14 *Show that the following limits do not exist:*

1. $\displaystyle \lim_{(x,y) \to (0,0)} \frac{x^2}{x^2 + y^3}$

2. $\displaystyle \lim_{(x,y) \to (0,0)} \frac{x^2 y}{x^3 + y^3}$

3. $\displaystyle \lim_{(x,y) \to (0,0)} \frac{x + y}{\sqrt{x^2 + y^2}}$

4. $\displaystyle \lim_{(x,y) \to (0,0)} \frac{x^2 y^2}{x^3 + y^3}$

2.2 Continuity

We say a function $f(x, y)$ is *continuous* at (a, b) if its limit as (x, y) approaches (a, b) equals its value there. In symbols we write

$$\lim_{(x,y)\to(a,b)} f(x, y) = f(a, b)$$

Most functions you can easily write down are continuous at every point of their domain. Hence, what you want to avoid are points outside of the domain, where you may have

1. Division by zero.
2. Square roots of negatives.
3. Logs of nonpositive numbers.
4. Tangents of odd multiples of $\frac{\pi}{2}$.

In each of these situations the function does not even exist, in which case it is certainly not continuous. But even if the function exists it may not have a limit. And even if the function exists, and the limits exist, they may not be equal.

EXAMPLE 2-4
Suppose

$$f(x, y) = \frac{x + y}{\sqrt{x^2 + y^2}}$$

There is no zero in the denominator when $(x, y) = (1, 1)$, so $f(x, y)$ is continuous at $(1, 1)$.

EXAMPLE 2-5
Evaluate

$$\lim_{(x,y)\to(0,0)} \frac{x^2 y^3}{x^2 + y^2 + 1}$$

There are no values of x and y that will make the denominator 0, so the function is continuous everywhere. Since the value of a continuous function equals its limit, we can evaluate the above simply by plugging in $(0, 0)$.

$$\lim_{(x,y)\to(0,0)} \frac{x^2 y^3}{x^2 + y^2 + 1} = \frac{0}{0 + 1} = 0$$

Problem 15 *Find the domain of the following functions:*

1. $\frac{x+y}{x-y}$
2. $\sqrt{y-x^2}$
3. $\ln(y-x)\sqrt{x-y}$

Problem 16 *Consider the function*

$$f(x,y) = \begin{cases} \frac{x^2+y^2}{\sin(x^2+y^2)} & (x,y) \neq (0,0) \\ 1 & (x,y) = (0,0) \end{cases}$$

Is $f(x,y)$ continuous at $(0,0)$?

Quiz

Problem 17

1. *Show that the function*

$$f(x,y) = \frac{x\sin y}{x^2+y^2}$$

   does not have a limit as $(x,y) \to (0,0)$.
2. *Is the function*

$$f(x,y) = \begin{cases} \frac{x+y}{x+y} & (x,y) \neq (0,0) \\ 1 & (x,y) = (0,0) \end{cases}$$

   continuous at $(0,0)$?
3. *Find the domain of the function*

$$f(x,y) = \ln\frac{1}{x-y^2}.$$

CHAPTER 3

Derivatives

3.1 Partial Derivatives

What shall we mean by the *derivative* of $f(x, y)$ at a point (x_0, y_0)? Just as in one variable calculus, the answer is the slope of a tangent line. The problem with this is that there are multiple tangent lines one can draw to the graph of $z = f(x, y)$ at any given point. Which one shall we pick to represent the derivative? The answer is another question: "*Which* derivative?" We will see that at any given point there are lots of possible derivatives; one for each tangent line.

Another way to think about this is as follows. Suppose we are at the point (x_0, y_0) and we start moving. While we do this we keep track of the quantity $f(x, y)$. The rate of change that we observe is the derivative, but the answer may depend on which direction we are traveling.

Suppose, for example, that we are observing the function $f(x, y) = x^2 y$, while moving through the point $(1, 1)$ with unit speed. Suppose further that we are traveling parallel to the x-axis, so that our y-coordinate is always one. We would like to know the observed rate of change of $f(x, y)$. Since the y-coordinate is always one the values of $f(x, y)$ that we observe are always determined by our x-coordinate: $f(x, 1) = x^2$. The rate of change of this function is given by its derivative: $2x$. Finally, when $x = 1$ this is the number 2.

The above is a particularly easy computation. Given any function, if you are traveling in a direction which is parallel to the x-axis then your y-coordinate is fixed. Plugging this number in for y then gives a function of just x, which we can differentiate. Here's another example.

EXAMPLE 3-1
We compute the rate of change of $f(x, y) = x^3 y^3$ at the point $(1, 2)$, when we are traveling parallel to the x-axis. During our travels the value of y stays fixed at 2. Hence, the values of the function we are observing are determined by our x-coordinate: $f(x, 2) = 8x^3$. The derivative of this function is then $24x^2$, which takes on the value 24 when x is one.

What if we wanted to repeat our computations, with different values of y? It would be helpful to keep the letter "y" in our computations, and plug in the value at the very end. Notice that when we plugged in a number for y it became a constant, and was treated as such when we differentiated with respect to x. If we leave the letter y in our computations we can still treat it as a constant.

EXAMPLE 3-2
Let $f(x, y) = x + xy + y^2$. We wish to treat y as a constant, just as if we had plugged in a number for it, and take the derivative with respect to x. Recall that the derivative of a sum of functions is the sum of the derivatives. So we will discuss the derivatives of each of the terms of $x + xy + y^2$ individually.

There is no occurrence of y in the first term, so it is particularly easy. Its derivative is just one.

The second term is a little trickier. Since we are treating y as a constant this is of the form $const \cdot x$. The derivative of such a function is just $const$. So the derivative of xy is just y.

Finally, the quantity y^2 is also a constant. The derivative of a constant is zero. Hence our answer is $1 + y$.

Notice in the above example that if we thought of x as constant, and y as the variable, then the derivative would have been very different. We need some notation to tell us what is changing and what is being kept constant. We use the symbols $\frac{\partial f}{\partial x}$ to represent the *partial derivative with respect to* x. This means x is considered a

variable, every other letter is a constant, and we differentiate. Similarly, the notation $\frac{\partial f}{\partial y}$ is the *partial derivative with respect to y*.

EXAMPLE 3-3

Again we let $f(x, y) = x + xy + y^2$. We compute

$$\frac{\partial f}{\partial y} = x + 2y$$

There are various other ways to think of partial derivatives that are useful. One is completely algebraic. Recall that the derivative with respect to x of a function $f(x)$ of one variable is defined as the limit

$$\frac{df}{dx} = \lim_{\Delta x \to 0} \frac{f(x + \Delta x) - f(x)}{\Delta x}$$

The partial with respect to x is defined similarly. Just remember that y is kept constant, so that $f(x, y)$ really becomes a function of just x. We then apply the above definition to get:

$$\frac{\partial f}{\partial x} = \lim_{\Delta x \to 0} \frac{f(x + \Delta x, y) - f(x, y)}{\Delta x}$$

The partial with respect to y is defined similarly

$$\frac{\partial f}{\partial y} = \lim_{\Delta y \to 0} \frac{f(x, y + \Delta y) - f(x, y)}{\Delta y}$$

There is yet another way to think about the partial derivative. We began this section by claiming that the derivative will still represent the slope of a tangent line. In the following figure we see the graph of an equation $z = f(x, y)$. Through the point (x_0, y_0) in the xy-plane there is also drawn a vertical plane parallel to the xz-plane. The intersection of this vertical plane with the graph is a curve. The slope of the tangent line to this curve is exactly the value of the partial derivative with respect to x at (x_0, y_0). To see the partial derivative with respect to y we would have a similar picture, where the vertical plane is parallel to the yz-plane.

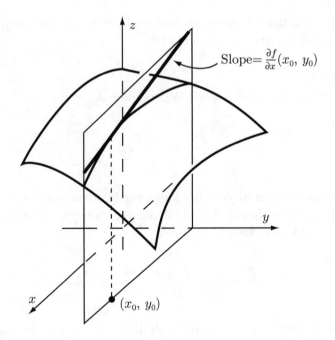

Problem 18 *Compute $\frac{\partial f}{\partial x}(2, 3)$ and $\frac{\partial f}{\partial y}(2, 3)$ for the following functions.*

1. $x + xy$
2. $x \ln y$
3. $x\sqrt{xy}$

Problem 19 *Compute $\frac{\partial f}{\partial x}(x, y)$ and $\frac{\partial f}{\partial y}(x, y)$ for the following functions.*

1. $x^2 y^3$
2. $\frac{x}{y}$

Problem 20 *For the function $f(x, y) = -x + xy^2 - y^2$ find all places where both $\frac{\partial f}{\partial x}$ and $\frac{\partial f}{\partial y}$ are zero.*

3.2 Composition and the Chain Rule

3.2.1 COMPOSITION WITH PARAMETERIZED CURVES

Suppose we have a parameterized curve $\phi(t) = (x(t), y(t))$ in the plane. That is, for a given value of t we are given the numbers $x(t)$ and $y(t)$, which we visualize as a point in the plane. We can also take these two numbers and plug them in to

a function $f(x, y)$. The result is the composition $f(\phi(t))$. Notice that only one number goes in to this function, and only one number comes out.

EXAMPLE 3-4
Let $f(x, y) = x^2 + y^2$. Let $\phi(t) = (t \cos t, t \sin t)$. Then

$$f(\phi(t)) = f(t \cos t, t \sin t) = t^2 \cos^2 t + t^2 \sin^2 t = t^2$$

There are various ways to visualize the composition. One is to imagine the graph of $z = f(x, y)$ in three dimensions. Then draw the parameterized curve $\phi(t)$ in the xy-plane, and imagine a vertical piece of paper curled up so that it sits on this curve. Now mark where the paper intersects the graph of $f(x, y)$, and unroll it. The result is the graph of $f(\phi(t))$.

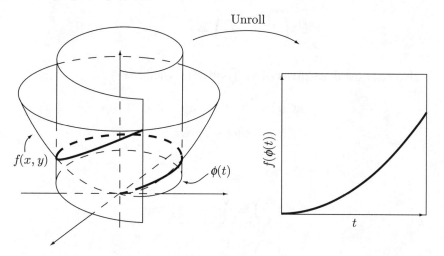

Since $f(\phi(t))$ is a function of one variable we can talk about its derivative just as if we were in a first term calculus class. But what we want to do here is relate it to the derivatives of $f(x, y)$, $x(t)$, and $y(t)$. The formula we end up with is the multivariable calculus version of the "chain rule"

$$\frac{d}{dt} f(\phi(t)) = \frac{\partial f}{\partial x} \frac{dx}{dt} + \frac{\partial f}{\partial y} \frac{dy}{dt}$$

EXAMPLE 3-5
Let $f(x, y)$ and $\phi(t)$ be defined as in the previous example. We wish to compute $\frac{d}{dt} f(\phi(t))$ when $t = \frac{\pi}{6}$. Note first that

$$\phi\left(\frac{\pi}{6}\right) = \left(\frac{\sqrt{3}\pi}{12}, \frac{\pi}{12}\right)$$

To use the chain rule we must compute the partial derivatives of $f(x, y)$ at this point. First note that

$$\frac{\partial f}{\partial x}(x, y) = 2x, \text{ and}$$

$$\frac{\partial f}{\partial y}(x, y) = 2y$$

Thus,

$$\frac{\partial f}{\partial x}\left(\frac{\sqrt{3}\pi}{12}, \frac{\pi}{12}\right) = \frac{\sqrt{3}\pi}{6}, \text{ and}$$

$$\frac{\partial f}{\partial y}\left(\frac{\sqrt{3}\pi}{12}, \frac{\pi}{12}\right) = \frac{\pi}{6}$$

We also need the derivatives of $x(t)$ and $y(t)$ when $t = \frac{\pi}{6}$.

$$x'(t) = \cos t - t \sin t, \text{ and}$$

$$y'(t) = \sin t + t \cos t$$

Thus,

$$x'\left(\frac{\pi}{6}\right) = \frac{\sqrt{3}}{2} - \frac{\pi}{12}, \text{ and}$$

$$y'\left(\frac{\pi}{6}\right) = \frac{1}{2} + \frac{\sqrt{3}\pi}{12}$$

Finally, we have

$$\frac{d}{dt}f\left(\phi\left(\frac{\pi}{6}\right)\right) = \frac{\partial f}{\partial x}\frac{dx}{dt} + \frac{\partial f}{\partial y}\frac{dy}{dt}$$

$$= \frac{\sqrt{3}\pi}{6}\left(\frac{\sqrt{3}}{2} - \frac{\pi}{12}\right) + \frac{\pi}{6}\left(\frac{1}{2} + \frac{\sqrt{3}\pi}{12}\right)$$

$$= \frac{\pi}{3}$$

Now that we have gone through the pain of using the chain rule to compute the derivative, it should be pointed out that the same answer could have been found *much* faster directly. We know from the previous example that $f(\phi(t)) = t^2$. So the derivative function is $2t$. Plugging in $t = \frac{\pi}{6}$ then immediately gives $(2)\left(\frac{\pi}{6}\right) = \frac{\pi}{3}$.

Problem 21 *Let $\phi(t) = (2t, t^2)$. Suppose you don't know what $f(x, y)$ is, but you know $\frac{\partial f}{\partial x}(2, 1) = 6$ and $\frac{\partial f}{\partial y}(2, 1) = -1$. Compute $\frac{d}{dt} f(\phi(1))$.*

Problem 22 *Suppose you don't know what $\phi(t) = (x(t), y(t))$ is, but you know $\phi(2) = (1, 3)$, $x'(2) = -2$, and $y'(2) = 1$. Let $f(x, y) = x^2 y$. Compute $\frac{d}{dt} f(\phi(2))$.*

3.2.2 COMPOSITION OF FUNCTIONS OF MULTIPLE VARIABLES

We now look at the idea of composition with more complicated types of functions, as in our next example.

EXAMPLE 3-6
Let $x(u, v) = uv$ and $y(u, v) = u^2 + v^2$. Suppose $f(x, y) = x + y$. Then we may form the composition $f(x(u, v), y(u, v))$ as follows.

$$f(x(u, v), y(u, v)) = x(u, v) + y(u, v) = uv + u^2 + v^2$$

Notice that the result is a function whose input is a pair of numbers, u and v, and whose output is a single number. Hence, we may talk about the partial derivatives, with respect to u and v, of the function given by composition. The result is given by another variant of the chain rule. As in the previous example, we will begin with the functions $x(u, v)$, $y(u, v)$, and $f(x, y)$. The following formulas give the partial derivatives of the composition $f(x(u, v), y(u, v))$.

$$\frac{\partial f}{\partial u} = \frac{\partial f}{\partial x}\frac{\partial x}{\partial u} + \frac{\partial f}{\partial y}\frac{\partial y}{\partial u}$$

$$\frac{\partial f}{\partial v} = \frac{\partial f}{\partial x}\frac{\partial x}{\partial v} + \frac{\partial f}{\partial y}\frac{\partial y}{\partial v}$$

At this point it is mathematically meaningless to think of terms like "∂x" as entities in themselves that can be canceled, but thinking this way may help you remember the above formulas. Note that the formula for the partial derivative of f with respect to u, for example, has two terms. If you cancel the "quantity" ∂x in the first term, and ∂y in the second, you are left with $\frac{\partial f}{\partial u}$ in both.

EXAMPLE 3-7

Let $f(x, y) = x^2 + xy$. Suppose we don't know $x(u, v)$ or $y(u, v)$, but we know $x(1, 2) = 3$, $y(1, 3) = -2$, $\frac{\partial x}{\partial u}(1, 2) = -1$, and $\frac{\partial y}{\partial u}(1, 2) = 5$. Then we may use the chain rule to compute $\frac{\partial f}{\partial u}(1, 2)$. To do this we will need to know $\frac{\partial f}{\partial x}$ and $\frac{\partial f}{\partial y}$ at the point $(x(1, 2), y(1, 2)) = (3, -2)$.

$$\frac{\partial f}{\partial x}(3, -2) = (2)(3) + (-2) = 4, \quad \frac{\partial f}{\partial y}(3, -2) = 3$$

We now employ the chain rule:

$$\frac{\partial f}{\partial u} = \frac{\partial f}{\partial x}\frac{\partial x}{\partial u} + \frac{\partial f}{\partial y}\frac{\partial y}{\partial u}$$

$$= (4)(-1) + (3)(5)$$

$$= 11$$

Problem 23 *The following table lists values for a function $f(x, y)$ and its partial derivatives.*

(x, y)	$f(x, y)$	$\frac{\partial f}{\partial x}$	$\frac{\partial f}{\partial y}$
$(1, 1)$	-3	-2	2
$(1, 2)$	5	1	1
$(2, 1)$	2	0	7
$(2, 5)$	-1	2	3
$(2, 3)$	11	1	-1
$(3, 2)$	2	1	0

Let $x(u, v) = uv$ and $y(u, v) = u + v^2$. Find the following partial derivatives at the indicated points.

1. $\frac{\partial f}{\partial u}$ *at* $(u, v) = (1, 2)$
2. $\frac{\partial f}{\partial v}$ *at* $(u, v) = (2, 1)$

Problem 24 *Let $f(x, y) = \sin(x + y)$, $x(u, v) = u + v$, and $y(u, v) = u - v$. Find the following quantities when $(u, v) = (\frac{\pi}{2}, \pi)$.*

1. $f(x, y)$
2. $\frac{\partial f}{\partial u}$
3. $\frac{\partial f}{\partial v}$

3.3 Second Partials

In the previous sections we saw that the partial derivatives of a function $f(x, y)$ are also functions of x and y. We can therefore take the derivative again with respect to either variable.

EXAMPLE 3-8
Let $f(x, y) = x^2 y^3$. The partial derivatives are $\frac{\partial f}{\partial x} = 2xy^3$ and $\frac{\partial f}{\partial y} = 3x^2 y^2$. We can take the partial derivative again of both of these functions with respect to either variable:

$$\frac{\partial}{\partial x}\left(\frac{\partial f}{\partial x}\right) = 2y^3, \quad \frac{\partial}{\partial y}\left(\frac{\partial f}{\partial x}\right) = 6xy^2$$

$$\frac{\partial}{\partial x}\left(\frac{\partial f}{\partial y}\right) = 6xy^2, \quad \frac{\partial}{\partial y}\left(\frac{\partial f}{\partial y}\right) = 6x^2 y$$

As is customary, we adopt the following shorthand notations for the second derivatives:

$$\frac{\partial}{\partial x}\left(\frac{\partial f}{\partial x}\right) = \frac{\partial^2 f}{\partial x^2}, \quad \frac{\partial}{\partial y}\left(\frac{\partial f}{\partial x}\right) = \frac{\partial^2 f}{\partial y \partial x}$$

$$\frac{\partial}{\partial x}\left(\frac{\partial f}{\partial y}\right) = \frac{\partial^2 f}{\partial x \partial y}, \quad \frac{\partial}{\partial y}\left(\frac{\partial f}{\partial y}\right) = \frac{\partial^2 f}{\partial y^2}$$

The quantities $\frac{\partial^2 f}{\partial y \partial x}$ and $\frac{\partial^2 f}{\partial x \partial y}$ are called the *mixed partials*. The above example illustrates an amazing fact: Under reasonable conditions the mixed partials are always equal! The "reasonable conditions" are only that the mixed partials exist and are themselves continuous functions. The proof of this goes back to the limit definition of the partial derivative.

Problem 25 *Find all second partial derivatives of the following functions.*

1. $f(x, y) = xy$
2. $f(x, y) = x^2 - y^2$
3. $f(x, y) = \sin(xy^2)$

Quiz

Problem 26 *Let* $f(x, y) = x^2 y + x^3 y^2$.

1. *Find* $\frac{\partial f}{\partial x}$ *and* $\frac{\partial f}{\partial y}$.

2. *If* $\phi(t) = (t^2, t - 1)$, *then what is* $f(\phi(t))$?

3. *Suppose you don't know what* $\psi(t) = (x(t), y(t))$ *is, but you know* $\psi(2) = (1, 1)$, $\frac{dx}{dt}(2) = 3$, *and* $\frac{dy}{dt}(2) = 1$. *Find the derivative of* $f(\psi(t))$ *when* $t = 2$.

4. *Suppose* x *and* y *are functions of* u *and* v, $x(u, v) = u^2 + v$, *and* $y(1, 1) = 1$. *What would* $\frac{\partial y}{\partial u}$ *have to be when* $(u, v) = (1, 1)$, *if* $\frac{\partial f}{\partial u} = 12$?

CHAPTER 4

Integration

4.1 Integrals over Rectangular Domains

The integral of a function of one variable gives the area under the graph and above an interval on the x-axis called the *domain of integration*. For functions of two variables the graph is a surface. We will interpret the act of integration as that of finding the volume below the surface, and above some region in the xy-plane. Eventually, we will examine how to do this with very general-shaped regions. We begin by looking at rectangles.

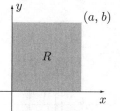

Let R be the region of the xy-plane pictured above. Suppose we want to find the volume of the region in $\mathbb{R}^3$ which sits above R, and below the graph of $z = f(x, y)$, as in the following figure. We do this by following these steps.

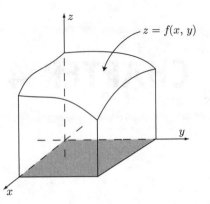

1. Begin by choosing a grid of points $\{(x_i, y_j)\}$ in R, so that the horizontal and vertical spacings between adjacent points are Δx and Δy, respectively.

2. Connect these grid points to break up R into rectangles. Note that the area of each rectangle is $\Delta x \Delta y$.

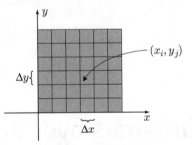

3. We now draw a box of height $f(x_i, y_j)$ above each such rectangle to get a figure which approximates the desired volume. The volume of each box is its length $\times$ width $\times$ height, which is $f(x_i, y_j)\Delta x \Delta y$.

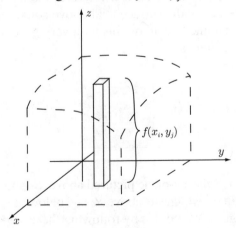

4. We now add up the volumes of all of these boxes to obtain the quantity

$$\sum_j \sum_i f(x_i, y_j) \Delta x \Delta y$$

5. Finally, we repeat this process indefinitely, each time using a grid with Δx and Δy smaller and smaller. The corresponding figures that we get approximate the desired volume more and more. In the limit we get what we're after, which we denote as $\int\int_R f(x, y)\, dx\, dy$:

$$\iint\limits_R f(x, y)\, dx\, dy = \lim_{\Delta y \to 0} \lim_{\Delta x \to 0} \sum_j \sum_i f(x_i, y_j) \Delta x \Delta y$$

Basic properties of summation and limits allow us to rearrange the above equation as follows:

$$\iint\limits_R f(x, y)\, dx\, dy = \lim_{\Delta y \to 0} \sum_j \left[\lim_{\Delta x \to 0} \sum_i f(x_i, y_j) \Delta x \right] \Delta y$$

The quantity in the brackets $\lim_{\Delta x \to 0} \sum_i f(x_i, y_j) \Delta x$ is exactly the definition of what you get from $f(x, y)$ when you fix y and integrate x, as in the following example.

EXAMPLE 4-1
Let $f(x, y) = x^2 + y^3$. If we fix $y = 3$ then this is the function $f(x) = x^2 + 27$. We may now integrate this function over some range of values of x, such as $[0, 2]$:

$$\lim_{\Delta x \to 0} \sum_i f(x_i, y_j) \Delta x = \int_0^2 x^2 + 27\, dx = \frac{1}{3}x^3 + 27x \Big|_{x=0}^{2} = \frac{8}{3} + 56$$

If we don't substitute the value 3 for y, but we still think of y as a constant, very little changes:

$$\int_0^2 x^2 + y^3\, dx = \frac{1}{3}x^3 + y^3 x \Big|_{x=0}^{2} = \frac{8}{3} + 2y^3$$

In the above example, notice that when we think of y as a constant and we integrate with respect to x, our answer is a function of y. We may now integrate this new function with respect to y. This is precisely what we are instructed to do in the limit we obtained for $\iint\limits_R f(x, y)\, dx\, dy$.

$$\iint\limits_R f(x, y)\, dx\, dy = \lim_{\Delta y \to 0} \sum_j \left[\lim_{\Delta x \to 0} \sum_i f(x_i, y_j)\Delta x \right] \Delta y$$

$$= \lim_{\Delta y \to 0} \sum_j \left[\int_0^a f(x, y)\, dx \right] \Delta y$$

$$= \int_0^b \left[\int_0^a f(x, y)\, dx \right] dy$$

EXAMPLE 4-2

Let $f(x, y) = xy^2$. Suppose Q is the rectangle with vertices at $(1, 1)$, $(2, 1)$, $(1, 4)$, and $(2, 4)$. To find the volume under the graph of f and above this rectangle we wish to compute

$$\iint\limits_Q xy^2\, dx\, dy = \int_1^4 \left[\int_1^2 xy^2\, dx \right] dy$$

We work inside the brackets first. To do this integral we must pretend y is a constant.

$$\int_1^2 xy^2\, dx = \frac{1}{2}x^2 y^2 \Big|_1^2 = 2y^2 - \frac{1}{2}y^2 = \frac{3}{2}y^2$$

We now have

$$\int_1^4 \left[\int_1^2 xy^2\, dx \right] dy = \int_1^4 \frac{3}{2}y^2\, dy = \frac{1}{2}y^3 \Big|_1^4 = 32 - \frac{1}{2} = \frac{63}{2}$$

Note that in the definition of $\iint\limits_R f(x, y)\, dx\, dy$ we could have rearranged the terms so that the integral with respect to y was inside the brackets. The fact that the answer does not depend on what order you integrate is called *Fubini's Theorem*.

EXAMPLE 4-3
Let S be the rectangle with vertices at $(1, 0)$, $(2, 0)$, $(1, 1)$, and $(2, 1)$. We integrate the function $f(x, y) = e^{\frac{y}{x}}$. We compute the integral of $f(x, y)$ over S by doing the integral with respect to y first:

$$\iint_S e^{\frac{y}{x}} \, dx \, dy = \int_1^2 \left[\int_0^1 e^{\frac{y}{x}} \, dy \right] dx$$

$$= \int_1^2 \left[x e^{\frac{y}{x}} \Big|_{y=0}^{1} \right] dx$$

$$= \int_1^2 \left[x e^{\frac{1}{x}} - x \right] dx$$

$$= -x e^{-x} - e^{-x} - \frac{1}{2} x^2 \Big|_1^2$$

$$= -\frac{3}{e^2} + \frac{2}{e} - \frac{3}{2}$$

To save space the integral $\int_a^b [\int_c^d f(x, y) dx] dy$ is often written as $\int_a^b \int_c^d f(x, y) \, dx \, dy$.

Problem 27 *Compute the following:*

1. $\int_0^1 \int_2^3 x + xy^2 \, dx \, dy$

2. $\int_{-1}^1 \int_0^1 x^2 y^2 \, dy \, dx$

3. $\int_0^{\frac{\pi}{2}} \int_0^{\frac{\pi}{2}} \cos(x + y) \, dx \, dy$

Problem 28 *Let R be the rectangle in the xy-plane with corners at $(-1, -1)$, $(-1, 0)$, $(2, -1)$, and $(2, 0)$. Find the volume below the graph of $z = x^3 y$ and above R.*

Problem 29 *Let R be the rectangle with vertices at* $(0, 0)$, $(1, 0)$, $(0, 1)$, *and* $(1, 1)$. *Find a formula for* $\int\int_R x^n y^m \, dx \, dy$.

Problem 30 *Let V be the volume below the graph of* e^{-xy^2}, *and above the rectangle with corners at* $(0, 0)$, $(2, 0)$, $(0, 3)$, *and* $(2, 3)$. *Find the area of the intersection of V with the vertical plane through the point* $(1, 1)$ *which is parallel to the xz-plane.*

4.2 Integrals over Nonrectangular Domains

In the previous section we saw how to compute the volume under the graph of a function and above a rectangle. But what if we are interested in the volume which lies above a nonrectangular area? For example, suppose now that R is the region of the xy-plane that is bounded by the graph of $y = g(x)$, the x-axis, and the line $x = 1$. We would like to find the volume of the figure V which lies below the graph of $z = f(x, y)$ and above the region R.

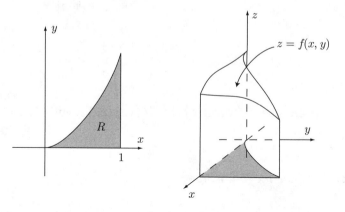

In the previous section we computed volume by breaking up the region in question into small boxes. Our strategy here is to cut it into "slabs" parallel to the yz-plane. To specify the location of each slab we must give a value of x. Hence, the volume of each slab, calculated as an integral with respect to y, is a function of x. Adding the slabs up is just like integrating with respect to x.

We follow these steps:

1. Choose points $\{x_i\}$ in the interval $[0, 1]$.

2. Compute the area $A(x_i)$ of the intersection of V with the plane $x = x_i$. We do this by plugging x_i in for the variable x in the function $f(x, y)$ and then integrating with respect to y. Notice that y ranges from 0 to $g(x_i)$ on

this domain. So the requisite area is given by

$$A(x_i) = \int_0^{g(x_i)} f(x_i, y)\, dy$$

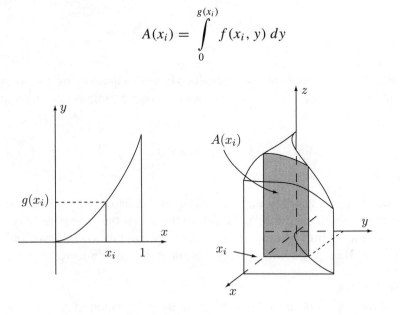

3. The volume of a thin slab is then given by

$$A(x_i)\Delta x = \int_0^{g(x_i)} f(x_i, y)\, dy\ \Delta x$$

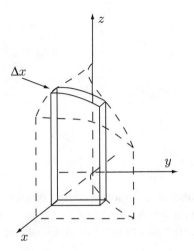

4. Adding up the volumes of all of the slabs thus gives

$$\sum_i \int_0^{g(x_i)} f(x_i, y) \, dy \, \Delta x$$

5. Finally, the desired volume is obtained from this quantity by choosing smaller and smaller slabs. This is equivalent to taking a limit as Δx tends toward 0.

$$V = \lim_{\Delta x \to 0} \sum_i \int_0^{g(x_i)} f(x_i, y) \, dy \, \Delta x = \int_0^1 \int_0^{g(x)} f(x, y) \, dy \, dx$$

The key to understanding this formula is the limits of integration. Each slab is parallel to the y-axis. The area of the side of the slab is thus computed as an integral with respect to y. But the range of values that y can take on depends on what the value of x is. Hence, the limits of integration of the inner integral depend on x.

EXAMPLE 4-4
Let R be the region in the xy-plane bounded by the graph of $y = x^2$, the x-axis, and the line $x = 1$. We compute the volume below the graph of $z = xy^2$ and above R as follows:

$$\text{Volume} = \int_0^1 \int_0^{x^2} xy^2 \, dy \, dx$$

$$= \int_0^1 \frac{1}{3} xy^3 \Big|_0^{x^2} dx$$

$$= \int_0^1 x^7 \, dx$$

$$= \frac{1}{8}$$

EXAMPLE 4-5
We now use the above ideas to compute a more complicated volume. Let Q be the region of the xy-plane bounded by the graphs of $y = x^2$ and $y = 1 - x^2$. We wish to determine the volume that lies below the graph of $z = x^3 + y$ and above Q.

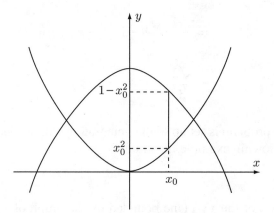

The above figure depicts the region Q. Notice that for a fixed value x_0 of x the values of y range from x^2 to $1 - x^2$. This tells us the limits of integration for the inner integral. Now notice that the smallest and largest possible values of x are where the graphs of x^2 and $1 - x^2$ coincide. To find this we solve

$$x^2 = 1 - x^2$$

Which implies

$$2x^2 = 1$$

And hence,

$$x = \pm \frac{\sqrt{2}}{2}$$

This tells us the limits of integration for the outer integral. We now compute the desired volume

$$\text{Volume} = \int_{-\frac{\sqrt{2}}{2}}^{\frac{\sqrt{2}}{2}} \int_{x^2}^{1-x^2} x^3 + y \, dy \, dx$$

$$= \int_{-\frac{\sqrt{2}}{2}}^{\frac{\sqrt{2}}{2}} x^3 y + \frac{1}{2} y^2 \Big|_{x^2}^{1-x^2} dx$$

$$= \int_{-\frac{\sqrt{2}}{2}}^{\frac{\sqrt{2}}{2}} -2x^5 + x^3 - x^2 + \frac{1}{2} \, dx$$

$$= -\frac{1}{3}x^6 + \frac{1}{4}x^4 - \frac{1}{3}x^3 + \frac{x}{2}\Big|_{-\frac{\sqrt{2}}{2}}^{\frac{\sqrt{2}}{2}}$$

$$= \frac{\sqrt{2}}{3}$$

Sometimes the problem is set up so that integrating with respect to x first is more natural, as the following example illustrates.

EXAMPLE 4-6
Let S be the region of the xy-plane bounded by the graph of $y = x^2$, the y-axis, and the lines $y = 1$ and $y = 2$. We will integrate the function $f(x, y) = x$ over this domain.

The domain S is pictured in the following figure. Notice that for a fixed value y_0 of y the range of values that x can take on goes from 0 to $\sqrt{y_0}$. This tells us the limits of integration for the inner integral. The smallest value y_0 can be is 1 and the largest value is 2. These are the limits for the outer integral.

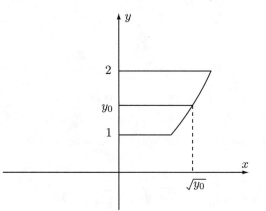

We now compute

$$\int_1^2 \int_0^{\sqrt{y}} x \, dx \, dy = \int_1^2 \frac{1}{2}x^2 \Big|_0^{\sqrt{y}} dy$$

$$= \int_1^2 \frac{1}{2}y \, dy$$

$$= \frac{1}{4}y^2 \Big|_1^2$$

$$= \frac{3}{4}$$

Often you can set up an integral in two ways: one so that dx comes first, and one so that dy comes first. Sometimes you'll find that only one of these ways makes the problem accessible. We illustrate this in the next example.

EXAMPLE 4-7
Let P be the region of the xy-plane bounded by the graph of $y = x$, the x-axis, and the line $x = 1$. We wish to integrate the function $f(x, y) = e^{-x^2}$ over P.

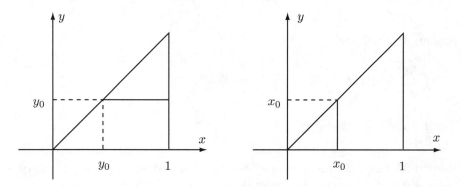

The region P is depicted above twice. Notice from the figure on the left that if we fix a value y_0 of y then x can range from y_0 to 1. The smallest possible value for y_0 is 0 and the biggest is 1. This tells us the limits of integration when we set up the integral with dx first.

$$\int_0^1 \int_y^1 e^{-x^2} \, dx \, dy$$

If, on the other hand, we fix a value x_0 of x then y can range from 0 to x, as in the figure on the right. The smallest possibility for x is 0 and the largest possibility is 1. This tells us that the integral can also be set up as

$$\int_0^1 \int_0^x e^{-x^2} \, dy \, dx$$

It is only possible to do the second of these integrals, which we compute as follows:

$$\int_0^1 \int_0^x e^{-x^2} \, dy \, dx = \int_0^1 ye^{-x^2}\Big|_0^x \, dx$$

$$= \int_0^1 xe^{-x^2} \, dx$$

$$= -\frac{1}{2}e^{-x^2}\Big|_0^1 \quad \text{(by } u\text{-substitution)}$$

$$= \frac{1}{2}\left(1 - \frac{1}{e}\right)$$

Problem 31 *Evaluate:*

1. $\int_0^1 \int_0^{y^2} 2xy^3 \, dx \, dy$

2. $\int_0^2 \int_x^{2x} e^{x+y} \, dy \, dx$

Problem 32 *Let T be the region of the xy-plane bounded by the graph of $y = x^2 - x - 2$ and the x-axis. Integrate the function $f(x, y) = x^2$ over T.*

Problem 33 *Let R be the region of the positive quadrant of the xy-plane bounded by $y = x$ and $y = x^3$. Set up two different integrals for $f(x, y)$ over R.*

Problem 34 *Evaluate the following integral by switching the order of integration:*

$$\int_0^{\pi^2} \int_{\sqrt{x}}^{\pi} \sin(y^3) \, dy \, dx$$

4.3 Computing Volume with Triple Integrals

The cylindrical figure pictured below has a base with area A and has height 1. The volume is therefore the product of these quantities, namely $A \cdot 1 = A$. But the top surface of this figure is the graph of the equation $z = 1$. The integral of this function gives the volume below the graph, which we just found out was A. In other words, if you want to find the area of a region R of the xy-plane, you just need to compute the integral of the function $f(x, y) = 1$ over R:

$$\text{Area}(R) = \iint\limits_{R} 1 \, dx \, dy$$

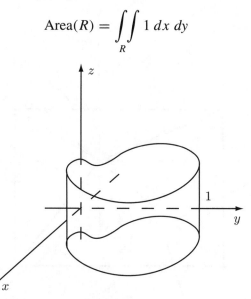

EXAMPLE 4-8
Suppose R is the region below the graph of $y = g(x)$, above the x-axis, and between the lines $x = a$ and $x = b$. Then we can find the area of R by evaluating the double integral:

$$\text{Area}(R) = \int\limits_{a}^{b} \int\limits_{0}^{g(x)} 1 \, dy \, dx$$

But this just reduces to

$$\int\limits_{a}^{b} y \Big|_{0}^{g(x)} \, dx = \int\limits_{a}^{b} g(x) \, dx$$

which should, of course, be familiar from first term calculus as the area under the graph of $y = g(x)$.

Just as one can compute area by evaluating the double integral of the function $f(x, y) = 1$, one can also compute volume by a *triple integral* of the function $f(x, y, z) = 1$. The tricky part usually involves finding the limits of integration.

Advanced Calculus Demystified

EXAMPLE 4-9

Let V be the region of $\mathbb{R}^3$ bounded by the planes $y = x$, $z = x + y$, $z = 0$, $y = 0$, and $x = 1$, as pictured below.

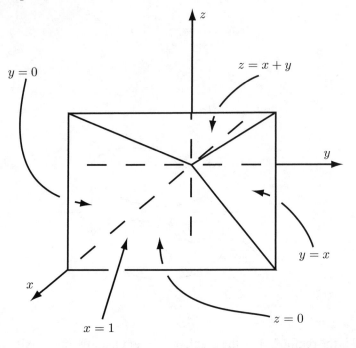

We set up a triple integral to compute the volume of V. The order of integration will be $dz\,dy\,dx$. To find the limits for the innermost integral (the one with respect to z) we fix x and y and observe that z can range from 0 to $x + y$. For the second integration we just fix x, and observe that y can vary from 0 to x. Finally, the smallest value that x can take on is 0 and the largest is 1, determining the limits of the outermost integral. We now compute

$$\text{Volume}(V) = \int_0^1 \int_0^x \int_0^{x+y} 1 \, dz \, dy \, dx$$

$$= \int_0^1 \int_0^x z\big|_0^{x+y} \, dy \, dx$$

$$= \int_0^1 \int_0^x x + y \, dy \, dx$$

$$= \int_0^1 \left. xy + \frac{1}{2}y^2 \right|_0^x dx$$

$$= \int_0^1 \frac{3}{2}x^2 \, dx$$

$$= \left. \frac{1}{2}x^3 \right|_0^1$$

$$= \frac{1}{2}$$

Problem 35 *Use a triple integral to find the volume of the solid bounded by the graph of $z = 1 - y^2$, and the planes $z = 0$, $x = 0$, and $x = 1$.*

Problem 36 *Find the volume of the solid bounded by the graph of $z = 1 - x - y^2$, and the coordinate planes $z = 0$ and $x = 0$.*

Problem 37 *Set up a triple integral to find the volume which lies below the paraboloid $z = 1 - x^2 - y^2$ and above the xy-plane.*

Problem 38 *Write a triple integral which will compute the volume bounded by the sphere $x^2 + y^2 + z^2 = 1$.*

QUIZ

Problem 39

1. *Evaluate the following integrals:*

    a. $\int_1^2 \int_2^3 \cos(2x + y) \, dx \, dy$

    b. $\int_0^1 \int_x^1 \sqrt{1 + y^2} \, dy \, dx$

2. *Set up an integral for the volume which lies between the cone $\sqrt{1 - x^2 - y^2}$ and the xy-plane.*

CHAPTER 5

Cylindrical and Spherical Coordinates

5.1 Cylindrical Coordinates

A *coordinate system* is a systematic way of locating a point in (some) space by specifying a few numbers. The coordinate system you are most familiar with is called *rectangular coordinates*. In $\mathbb{R}^3$ this works by specifying x, y, and z, which represent distances along the x-, y-, and z-axes, respectively.

Rectangular coordinates are extremely cumbersome when trying to define certain common shapes, such as cylinders or spheres. A second way to determine the location of a point is to adapt polar coordinates to three dimensions. In this case we give a location in the xy-plane by a value of r and θ, and then the height off of the xy-plane by a value of z.

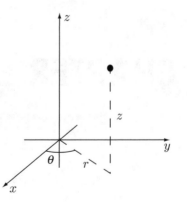

Converting from cylindrical coordinates to rectangular coordinates is as easy as it is in polar coordinates. The z coordinate is exactly the same in both systems. To get x and y you just use the formulas

$$x = r \cos \theta$$

$$y = r \sin \theta$$

EXAMPLE 5-1
We locate the point $r = 2$, $\theta = \frac{\pi}{6}$, and $z = 7$ in rectangular coordinates:

$$x = 2 \cos \frac{\pi}{6} = \sqrt{3}$$

$$y = 2 \sin \frac{\pi}{6} = 1$$

So the desired point is at $(\sqrt{3}, 1, 7)$.

Problem 40 *Write the rectangular coordinates of the following points.*

1. $(r, \theta, z) = (2, \frac{\pi}{4}, -1)$
2. $(r, \theta, z) = (0, \frac{\pi}{7}, 3)$
3. $(r, \theta, z) = (4, \frac{\pi}{3}, 0)$

5.2 Graphing Cylindrical Equations

To get a real feel for cylindrical coordinates it is helpful to look at the graphs of a few simple equations. The graph of something like $z = 3$ in cylindrical coordinates is the same as the graph of $z = 3$ in rectangular coordinates (it's the same z!): a horizontal plane that hits the z-axis at 3. The next two examples are a bit more interesting.

EXAMPLE 5-2
Consider the cylindrical equation $r = 3$. This is the set of all points in $\mathbb{R}^3$ which are exactly 3 units away from the z-axis. The shape is a cylinder of radius 3 centered on the z-axis.

EXAMPLE 5-3
We now examine the cylindrical equation $\theta = \frac{\pi}{4}$. This describes the set of all points that make an angle of $\frac{\pi}{4}$ with the xz-plane. The shape is a plane through the z-axis that is halfway between the xz-plane and the yz-plane.

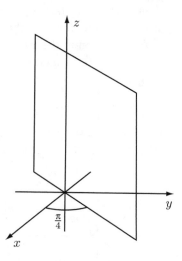

    More complicated equations involve relationships between r, θ, and z, as in the next example.

EXAMPLE 5-4
Consider the equation $z = r^2$. To visualize the graph we are being instructed to find all of the points in $\mathbb{R}^3$ whose z-coordinate is the square of their distance to the z-axis. The most relevant feature of the equation, however, is the lack of the coordinate θ. This tells us that no matter what θ is we will see the same picture.

Following this observation we consider the half plane which meets the z-axis at some arbitrary angle. Take this plane out of the picture and look at it head-on, as in the figure below. As you move left and right in this plane it is the quantity r that is changing. As you move up and down it is z that changes. Hence, it makes sense to label our axes r and z in this plane.

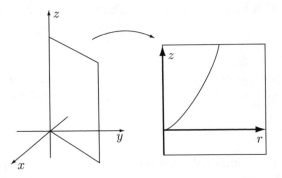

In the half plane at the right we have depicted the graph of $z = r^2$, a familiar parabola. Now, the key to understanding the complete picture is to place this parabola in *every* half plane at the left. The result is the surface that you get when you rotate the graph of $y = x^2$ (in $\mathbb{R}^2$) around the y-axis. It is called a *paraboloid*, and is depicted below (see also Figure 1-9 for a computer-generated picture).

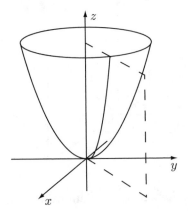

Problem 41 *Describe the shape of the following cylindrical equations.*

1. $\theta = 0$
2. $\theta = \frac{\pi}{2}$
3. $z = 0$

4. $z = r$

5. $z^2 + r^2 = 1$

Problem 42 *Sketch the graphs of the following equations.*

1. $z = \frac{1}{r}$

2. $r^2 - z^2 = 1$

3. $z^2 - r^2 = 1$

5.3 Spherical Coordinates

Cylindrical coordinates are very convenient for describing cylinders and other surfaces of revolution. In this section we explore *spherical coordinates*, which are useful for describing many other shapes.

To locate a point using spherical coordinates you need to know three quantities: ρ (pronounced "Row"), θ, and ϕ (pronounced "Fee"). The quantity θ is precisely the same as in cylindrical coordinates. ρ is the distance from the origin. ϕ is the angle with the z-axis, as pictured below.

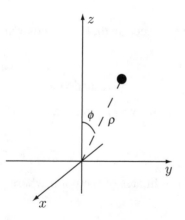

Given the spherical coordinates of a point we would like to be able to convert to rectangular coordinates. The easiest quantity to find is z. Consider the triangle in the figure below. The side adjacent to the angle ϕ is precisely z. The hypotenuse is ρ. So $\cos \phi = \frac{z}{\rho}$. Solving for z then gives

$$z = \rho \cos \phi$$

$$z = \rho \cos \phi$$

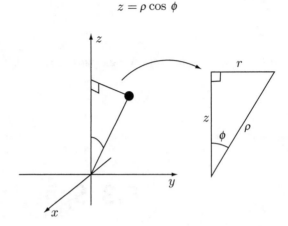

To find x and y it is helpful to first find the quantity r from cylindrical coordinates. Going back to the triangle in the above figure, we see that the side opposite the angle ϕ is precisely r. Hence, $\sin \phi = \frac{r}{\rho}$, and thus

$$r = \rho \sin \phi$$

We know from the previous section that $x = r \cos \theta$ and $y = r \sin \theta$. Substituting for r gives

$$x = \rho \sin \phi \cos \theta$$
$$y = \rho \sin \phi \sin \theta$$

EXAMPLE 5-5
We find the rectangular coordinates of the point where $\rho = 2$, $\theta = \frac{\pi}{3}$, and $\phi = \frac{\pi}{6}$.

$$x = 2 \sin \frac{\pi}{6} \cos \frac{\pi}{3} = \frac{1}{2}$$
$$y = 2 \sin \frac{\pi}{6} \sin \frac{\pi}{3} = \frac{\sqrt{3}}{2}$$
$$z = 2 \cos \frac{\pi}{6} = \sqrt{3}$$

Problem 43 *Convert the following to rectangular coordinates.*

1. $(\rho, \theta, \phi) = (0, \frac{\pi}{2}, \frac{\pi}{6})$
2. $(\rho, \theta, \phi) = (3, \pi, \frac{\pi}{2})$
3. $(\rho, \theta, \phi) = (4, \frac{\pi}{4}, \frac{\pi}{3})$

5.4 Graphing Spherical Equations

We now explore the graphs of several equations in spherical coordinates.

EXAMPLE 5-6

The simplest equation to understand is something like $\rho = 2$. This describes the set of all points that are precisely 2 units away from the origin. In other words, this is a sphere of radius 2.

If θ is a constant then the equation is the same in spherical coordinates and cylindrical coordinates; a plane through the z-axis. The previous example shows that when ρ is constant we get a sphere. In the next example we see what happens when ϕ is held constant.

EXAMPLE 5-7

Consider the equation $\phi = \frac{\pi}{6}$. This describes the points which make an angle of $\frac{\pi}{6}$ with the z-axis. This is a circular cone, centered on the z-axis.

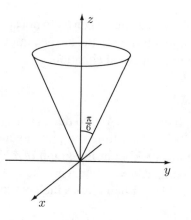

As with cylindrical coordinates, things get more difficult when equations involve relationships between multiple coordinates.

EXAMPLE 5-8
We graph the equation $\rho = \phi$. Notice that the coordinate θ does not appear in the equation. This tells us that the intersection of the graph with every half plane incident to the z-axis is the same. (Recall that we encountered a similar type of graph when we looked at some cylindrical equations.) This, in turn, tells us that the graph is a surface of revolution.

In the following figure we have sketched one half plane. In this half plane we have indicated how a spiral is the graph of $\rho = \phi$.

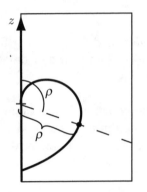

To form the complete graph we must now take this spiral and rotate it about the z-axis.

In the previous example we saw that when θ did not appear in a spherical equation the graph was a surface of revolution. In the next example we explore what happens when ϕ does not appear in a spherical equation.

EXAMPLE 5-9
We now explore the graph of $\rho = \theta$. We do this again by investigating various half planes incident to the z-axis. This time the pictures are not all the same. When $\theta = 0$, for example, we are looking at the graph of $\rho = 0$, which is just a point. When $\theta = \frac{\pi}{2}$ we are looking at $\rho = \frac{\pi}{2}$, which is half of a circle of radius $\frac{\pi}{2}$. As θ increases we see larger and larger half circles incident to the z-axis, as in the figure below.

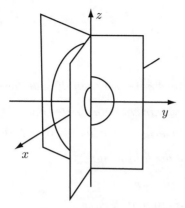

Putting it all together is challenging. We have done this for you in the computer-generated picture below. The result is reminiscent of a sea shell.

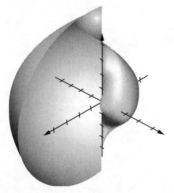

Problem 44 *Describe the graphs of the following spherical equations.*

1. $\rho = 0$
2. $\phi = \frac{\pi}{2}$
3. $\phi = \pi$
4. $\rho \sin \phi = 2$
5. $\rho \cos \phi = 2$

Problem 45 *Sketch the graphs of the following spherical equations.*

1. $\rho = \sin \phi$
2. $\rho = \cos \theta$

Quiz

Problem 46

1. *Give rectangular coordinates for the following points:*

   a. *The point with cylindrical coordinates $r = 1$, $\theta = \frac{\pi}{6}$, and $z = 2$.*

   b. *The point with spherical coordinates $\rho = 2$, $\theta = \frac{\pi}{6}$, and $\phi = \frac{\pi}{4}$.*

2. *Sketch the graphs of the following:*

   a. *The cylindrical equation $r = \cos\theta$.*

   b. *The spherical equation $\theta = \phi$.*

CHAPTER 6

Parameterizations

6.1 Parameterized Surfaces

In Chapter 1 we encountered parameterized curves. These were functions that had one input (the *parameter*) and multiple outputs. One visualizes such curves by drawing the range of the function. In this section we introduce *parameterized surfaces*. These are functions which have two inputs and multiple (usually three) outputs. Again, we visualize such functions by picturing their range. This is most often some surface in $\mathbb{R}^3$.

EXAMPLE 6-1
We explore the parameterization given by

$$\psi(u, v) = (u, v, u^2 + v^2)$$

Note that for each point in the range of ψ the z-coordinate is obtained from the x and y coordinates by squaring them and adding. In other words, if a point is in the range of ψ then it is also on the graph of $z = x^2 + y^2$. This is the paraboloid depicted in Figure 1-9.

EXAMPLE 6-2

We now look at a slight modification of the previous example.

$$\psi(u, v) = (2u, v, u^2 + v^2)$$

Note that this parameterization is obtained from the previous one by doubling the x-coordinate of every point. This has the effect of stretching the graph in the x-direction. Intersections with horizontal planes will thus be ellipses, as opposed to circles as in the previous example.

Example 6-1 illustrated another important idea. Suppose we want to represent the graph of a function $z = f(x, y)$ as a parameterized surface. Then we may simply write

$$\psi(u, v) = (u, v, f(u, v))$$

Note that the z-coordinate of every point in the range is obtained from the x- and y-coordinates by plugging them into f. Hence, each point in the range is indeed on the graph of f.

It is not much more difficult to represent graphs in other coordinate systems as parameterized surfaces. The trick is to always translate to rectangular coordinates.

EXAMPLE 6-3

Suppose we want to write the graph of the cylindrical equation $z = r^2$ as a parameterized surface. First, we write down the translation from cylindrical to rectangular coordinates:

$$x = r \cos \theta$$
$$y = r \sin \theta$$
$$z = z$$

Now we substitute r^2 for z:

$$x = r \cos \theta$$
$$y = r \sin \theta$$
$$z = r^2$$

Hence, a parameterization may be written as follows:

$$\psi(r, \theta) = (r \cos \theta, r \sin \theta, r^2)$$

EXAMPLE 6-4

Suppose we want to find a parameterization for the graph of the spherical equation

$$\rho = \cos\phi$$

First we write down the translation from spherical to rectangular coordinates:

$$x = \rho \sin\phi \cos\theta$$
$$y = \rho \sin\phi \sin\theta$$
$$z = \rho \cos\phi$$

Now we substitute $\cos\phi$ for ρ:

$$x = \cos\phi \sin\phi \cos\theta$$
$$y = \cos\phi \sin\phi \sin\theta$$
$$z = \cos\phi \cos\phi$$

This gives us the parameterization

$$\psi(\theta, \phi) = (\cos\phi \sin\phi \cos\theta, \cos\phi \sin\phi \sin\theta, \cos\phi \cos\phi)$$

Problem 47 *Find parameterizations for the graphs of the following equations.*

1. $z = x^2$
2. $r = \theta^2$
3. $\rho = \theta^2$

Problem 48 *The following are parameterizations of the graphs of rectangular, cylindrical, or spherical equations. Find these equations.*

1. $\phi(\rho, \theta) = (\rho \sin\theta \cos\theta, \rho \sin^2\theta, \rho \cos\theta)$
2. $\phi(x, z) = (x, x + xz, z)$
3. $\phi(r, \theta) = (r \cos\theta, r \sin\theta, \sin r)$

6.2 The Importance of the Domain

When you are trying to specify a particular shape, the domain of the parameterization can be just as important as the function. We illustrate this with an example.

EXAMPLE 6-5

Suppose we want to parameterize the part of a sphere of radius 1 that lies in the first octant (i.e., the region of $\mathbb{R}^3$ where $x, y, z \geq 0$). The equation of a sphere is particularly simple in spherical coordinates, so we start here. Such an equation is $\rho = 1$. Writing down the translation to rectangular coordinates and substituting 1 for ρ then yields the parameterization

$$\psi(\theta, \phi) = (\sin\phi\cos\theta, \sin\phi\sin\theta, \cos\phi)$$

But how do we make sure we just get the part of the sphere we want? The answer is to restrict the values of θ and ϕ that can be plugged into ϕ:

$$0 \leq \theta \leq \frac{\pi}{2}, \quad 0 \leq \phi \leq \frac{\pi}{2}$$

These ranges constitute the domain of ϕ.

EXAMPLE 6-6

We parameterize the portion of the graph of $z = r^2$ that lies inside the cylinder $x^2 + y^2 = 1$. The equation whose graph we are interested in is given in cylindrical coordinates, so it is natural to start there. The restriction on the domain can also be naturally expressed in cylindrical coordinates, so we do not anticipate any problems. In Example 6-3, we found a parameterization for the desired surface:

$$\psi(r, \theta) = (r\cos\theta, r\sin\theta, r^2)$$

The points of $\mathbb{R}^3$ that lie inside the cylinder $x^2 + y^2 = 1$ all have $r \leq 1$. Hence, we restrict the domain of our parameterization to

$$0 \leq r \leq 1, \quad 0 \leq \theta \leq 2\pi$$

Problem 49 *The function*

$$\psi(\theta, \phi) = (\sin \phi \cos \theta, \sin \phi \sin \theta, \cos \phi)$$

parameterizes a sphere of radius 1. How would you restrict the values of θ and ϕ to just get the portion of the sphere where

1. $x, y \geq 0$
2. $x \leq 0, y \geq 0, z \leq 0$

Problem 50 *Describe the difference between the shapes with the following parameterizations.*

1. $\Psi(x, y) = (x, y, \sqrt{x^2 + y^2}); 0 \leq x \leq 1, 0 \leq y \leq 1$
2. $\Psi(r, \theta) = (r \cos \theta, r \sin \theta, r); 0 \leq r \leq 1, 0 \leq \theta \leq 2\pi$
3. $\Psi(r, \theta) = (r \cos \theta, r \sin \theta, r); 0 \leq r \leq 1, 0 \leq \theta \leq \pi$

6.3 This Stuff Can Be Hard!

Parameterizations are extremely powerful tools that can describe very complicated surfaces. But with such great power can come great difficulty. There may be no obvious way to look at a complicated parameterization and know what shape it describes. For example, consider the parameterization

$$\phi(t, u) = ((4 + \cos t + 2 \sin u) \cos(2u),$$
$$(4 + \cos t + 2 \sin u) \sin(2u), \sin t + 2 \cos u)$$
$$0 \leq t \leq 2\pi, \quad 0 \leq u \leq 2\pi$$

No expert would be able to look at an equation so complicated and immediately know what it looks like. This is where computers become an invaluable tool. Plugging this equation into a standard graphing program reveals the following beautiful picture.

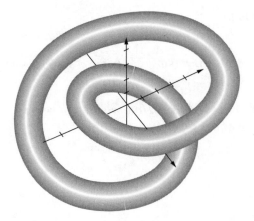

Despite the apparent difficulty in understanding such complicated functions, there are a few tricks that can help. These are basic ways in which you can take a parameterization and modify it to suit your needs. For example, if you multiply some coordinate by a constant, this has the effect of stretching the parameterized shape by that factor. If you add a constant to some coordinate the shape is moved in that direction.

EXAMPLE 6-7
Consider the following parameterization:

$$\Psi(\theta, \phi) = (2 \sin \phi \cos \theta + 1, 3 \sin \phi \sin \theta - 2, \cos \phi)$$

$$0 \leq \theta \leq 2\pi, \quad 0 \leq \phi \leq \pi$$

This looks much like a parameterization of a sphere from spherical coordinates. But the x-coordinate was multiplied by 2, and then 1 was added to it. Similarly, the y-coordinate was multiplied by 3 and then 2 was subtracted from it. Thus the parameterized shape is an ellipsoid with center at $(1, -2, 0)$.

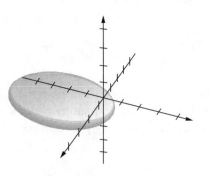

Problem 51 *Describe how the following parameterizations are related to graphs of functions in rectangular, cylindrical, or spherical coordinates.*

1. $\Psi(y, z) = (y^2 + z, 2y, z - 1)$
2. $\Psi(x, y) = (y, x, x + \sin y)$
3. $\Psi(\theta, z) = (2\cos\theta, 3\sin\theta + 1, z - 1)$

Problem 52 *Parameterize the surface whose level curves are the ellipses shown below.*

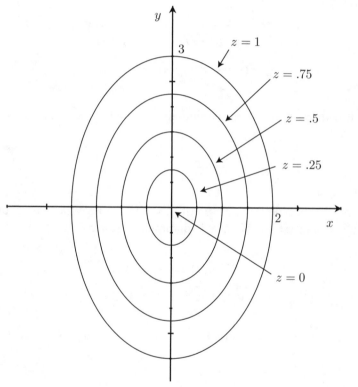

6.4 Parameterized Areas and Volumes

Earlier in the text we looked at parameterized curves in $\mathbb{R}^2$ and $\mathbb{R}^3$. This chapter began with parameterized surfaces in $\mathbb{R}^3$. Now we look at parameterized areas in $\mathbb{R}^2$ and parameterized volumes in $\mathbb{R}^3$. Often these types of parameterizations begin with writing down the translation from some coordinate system to rectangular coordinates, and then restricting the domain.

EXAMPLE 6-8
The translation from polar coordinates to rectangular coordinates is given by

$$x = r \cos \theta$$
$$y = r \sin \theta$$

We can use this to parameterize the area that is both inside the unit circle and in the first quadrant:

$$\Psi(r, \theta) = (r \cos \theta, r \sin \theta)$$
$$0 \leq r \leq 1, \quad 0 \leq \theta \leq \frac{\pi}{2}$$

Note that it is the restriction on the domain that guarantees that the correct area is parameterized.

EXAMPLE 6-9
Let R be the region below the graph of $y = f(x)$, above the x-axis, and between the lines $x = a$ and $x = b$. Then R is parameterized by

$$\Psi(x, t) = (x, t f(x))$$
$$a \leq x \leq b, \quad 0 \leq t \leq 1$$

In three dimensions one usually begins with rectangular, cylindrical, or spherical coordinates.

EXAMPLE 6-10
Consider the parameterization

$$\Psi(r, \theta, z) = (r \cos \theta, r \sin \theta, z)$$
$$1 \leq r \leq 2, \quad 0 \leq \theta \leq 2\pi, \quad 0 \leq z \leq 1$$

The function itself is just the translation from cylindrical to rectangular coordinates. The restriction on the domain means that we have only parameterized the region between cylinders of radii 1 and 2, and between the planes $z = 0$ and $z = 1$.

Of course, as in the previous section, one can distort and move shapes by multiplying and adding constants to each coordinate.

EXAMPLE 6-11
We examine the parameterization

$$\Psi(\rho, \theta, \phi) = (2\rho \sin\phi \cos\theta, \rho \sin\phi \sin\theta, \rho \cos\phi)$$

$$0 \le \rho \le 1, \quad 0 \le \theta \le \frac{\pi}{2}, \quad 0 \le \phi \le \frac{\pi}{2}$$

Without the 2 in the first component of the parameterization this would be one-eighth of a solid ball. The 2 just stretches it in the x-direction, making it one-eighth of an ellipsoid.

Problem 53 *Sketch the areas of the plane parameterized by the following.*

1. $\Psi(r, \theta) = (r \cos\theta, r \sin\theta), 0 \le r \le 1, 0 \le \theta, \le \dfrac{\pi}{4}$
2. $\Psi(r, \theta) = (r \cos\theta, 2r \sin\theta), 1 \le r \le 2, 0 \le \theta, \le \pi$

Problem 54 *Describe the volumes given by the following parameterizations.*

1. $\Psi(\rho, \theta, \phi) = (\rho \sin\phi \cos\theta, \rho \sin\phi \sin\theta, \rho \cos\phi), 1 \le \rho \le 2, 0 \le \theta \le 2\pi,$
   $0 \le \phi \le \frac{\pi}{2}$
2. $\Psi(r, \theta, z) = (r \cos\theta, r \sin\theta, 2z + 1), 0 \le r \le 1, 0 \le \theta \le 2\pi, 0 \le z \le 1$

Problem 55 *Find a parameterization for the volume which lies below the cone $z = r$, above the xy-plane, and inside the cylinder $x^2 + y^2 = 1$.*

Quiz

Problem 56

1. *Find a parameterization for the portion of the cylinder of radius 2, centered on the z-axis, which lies below the graph of $z = 2r$ and above the xy-plane.*

2. *Sketch the region in $\mathbb{R}^3$ parameterized by the following:*

$$\psi(\rho, \theta, \phi) = (2\rho \sin \phi \cos \theta, \rho \sin \phi \sin \theta, \rho \cos \phi)$$

where $0 \leq \rho \leq 1$, $0 \leq \theta \leq \frac{\pi}{2}$, and $0 \leq \phi \leq \pi$.

CHAPTER 7

Vectors and Gradients

7.1 Introduction to Vectors

Mathematically, there is very little difference between a vector and a point. In two dimensions, for example, a point is really just a pair of numbers. We visualize this pair as a dot in a plane. A two-dimensional vector is similar. Technically, it is also just a pair of numbers. But this time we visualize it as an arrow (Figure 7-1). This arrow does not have a preferred location. The two numbers that specify it only give its *magnitude* (i.e., size) and *direction*. This can be convenient, since certain algebraic operations with vectors can be seen visually by moving them around.

Some differences between points and vectors are purely cosmetic. For example, the numbers that you give to specify a point are called its *coordinates*. The numbers that specify a vector are its *components*. The main difference between points and vectors is that we can do algebra with vectors. This is a topic studied in detail in a class on *linear algebra*. We will need some of the ideas from linear algebra here.

The first algebraic operation we will study with vectors is addition. This is very easy: to add two vectors we simply add its components.

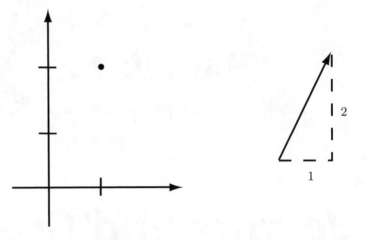

Figure 7-1   The point $(1, 2)$ and the vector $\langle 1, 2 \rangle$

EXAMPLE 7-1
We compute the sum of the vectors $\langle 2, 3 \rangle$ and $\langle -1, 2 \rangle$ as follows:

$$\langle 2, 3 \rangle + \langle -1, 2 \rangle = \langle 2 - 1, 3 + 2 \rangle = \langle 1, 5 \rangle$$

We can visualize the operation of addition as follows. To add a vector V to a vector W we slide W so that its "tail" (the base of the arrow) coincides with the "tip" of V. The result of $V + W$ is then the vector whose tail is at the tail of V and whose tip is at the tip of W.

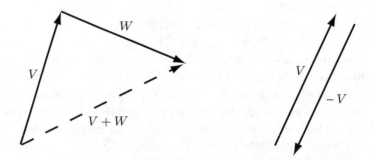

Negatives are equally easy. If V is a vector then we let $-V$ denote the vector whose components are the negatives of the components of V. The picture, of course, is a vector which has the same length as V, but points in the exact opposite direction.

EXAMPLE 7-2

If $V = \langle 2, 3 \rangle$ then $-V = \langle -2, -3 \rangle$.

Finally, we define subtraction by combining addition and negatives:

$$V - W = V + (-W)$$

The result of subtraction is therefore exactly what you would expect: the vector whose components are the difference between those of V and W.

The *magnitude* (or *length*) of a vector is easily computed by the Pythagorean Theorem. We denote the magnitude of a vector V by the absolute value bars $|\cdot|$. Hence,

$$|\langle a, b \rangle| = \sqrt{a^2 + b^2}$$

EXAMPLE 7-3

$$\left| \left\langle \frac{\sqrt{2}}{2}, \frac{\sqrt{2}}{2} \right\rangle \right| = \sqrt{\left(\frac{\sqrt{2}}{2} \right)^2 + \left(\frac{\sqrt{2}}{2} \right)^2} = \sqrt{\frac{1}{2} + \frac{1}{2}} = 1$$

A vector whose length is one is often called a *unit vector*.

Another common operation is multiplying a vector by a number. This is referred to as *scalar multiplication*. Like addition and subtraction, this is done component-wise. That is, the product of a scalar (i.e., a number) c and a vector V is a vector whose components are c times the components of V. That is,

$$c \langle a, b \rangle = \langle ca, cb \rangle$$

Geometrically, scalar multiplication produces a vector which points in the same direction, but whose magnitude has been altered by the given scale factor. For example, if we multiply a vector by the scalar 2, then we get a new vector, pointing in the same direction, which is twice as long.

EXAMPLE 7-4

Suppose we would like to find a unit vector that points in the same direction as $\langle 3, 4 \rangle$. The trick is to multiply it by some scalar, and arrange it so that the resulting vector has length one. In other words, we would like to find a number c such that

$$|c \langle 3, 4 \rangle| = 1$$

A quick computation gives us

$$|c\langle 3, 4\rangle| = |\langle 3c, 4c\rangle|$$
$$= \sqrt{(3c)^2 + (4c)^2}$$
$$= \sqrt{25c^2}$$
$$= 5c$$

Hence, $5c = 1$, and so $c = \frac{1}{5}$. We now multiply the vector $\langle 3, 4\rangle$ by this number to get the answer $\left\langle \frac{3}{5}, \frac{4}{5}\right\rangle$.

Problem 57 *For the following vectors V and W compute $V + W$, $-V$, and $V - W$.*

1. $V = \langle 1, 6\rangle$, $W = \langle 6, 1\rangle$
2. $V = \langle 0, 0\rangle$, $W = \langle 1, 2\rangle$
3. $V = \langle -1, 1\rangle$, $W = \langle 1, -2\rangle$

Problem 58 *Find the magnitude of the following vectors.*

1. $\langle 2, 3\rangle$
2. $\langle 1, 3\rangle$

Problem 59 *Find a unit vector that points in the same direction as $\langle 2, 1\rangle$.*

Problem 60 *Find a unit vector that is perpendicular to the vector $\langle 5, 12\rangle$.*

7.2 Dot Products

After we have mastered the basic operations of addition and subtraction we move on to multiplication. Unfortunately, we do not have anything quite as straightforward. It is tempting to define the product of V and W as the vector whose components are the products of the components of V and W. This particular operation, however, does not turn out to be terribly useful.

Instead we define the product of two vectors V and W to be the sum of the products of the components. This means the result of multiplication is a single number, as opposed to a vector. This product is called the *dot product*, since it is always denoted with a dot. In symbols, we write

$$\langle a, b\rangle \cdot \langle c, d\rangle = ac + bd$$

EXAMPLE 7-5

$$\langle 2, 3 \rangle \cdot \langle 3, 4 \rangle = 6 + 12 = 18$$

The usefulness of the dot product comes from the fact that it has a nice geometric interpretation. Suppose the angle between vectors V and W is θ. Then,

$$V \cdot W = |V||W|\cos\theta$$

This follows from some tricky trigonometry as follows. Consider the following triangle:

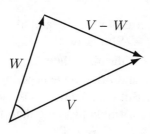

Now recall the *Law of Cosines*, which says that if a triangle has side lengths a, b, and c, and the angle between a and b is θ, then

$$c^2 = a^2 + b^2 - 2ab\cos\theta$$

If we apply this to the above triangle we get

$$|V - W|^2 = |V|^2 + |W|^2 - 2|V||W|\cos\theta$$

which we can rearrange to

$$|V||W|\cos\theta = \frac{|V|^2 + |W|^2 - |V - W|^2}{2}$$

Now suppose $V = \langle v_1, v_2 \rangle$ and $W = \langle w_1, w_2 \rangle$. Then,

$$\frac{|V|^2 + |W|^2 - |V - W|^2}{2} = \frac{v_1^2 + v_2^2 + w_1^2 + w_2^2 - (v_1 - w_1)^2 - (v_2 - w_2)^2}{2}$$

$$= \frac{2v_1 w_1 + 2v_2 w_2}{2}$$

$$= v_1 w_1 + v_2 w_2$$

And so

$$V \cdot W = v_1 w_1 + v_2 w_2 = |V||W| \cos \theta$$

In retrospect, this formula is very reasonable. Consider two extreme cases:

Case 1. Suppose we take the dot product of V with itself. By definition,

$$V \cdot V = v_1 v_1 + v_2 v_2 = v_1^2 + v_2^2$$

But in this case $\theta = 0$, so $\cos \theta = 1$. Hence,

$$|V||V| \cos \theta = |V||V| = \sqrt{v_1^2 + v_2^2}\sqrt{v_1^2 + v_2^2} = v_1^2 + v_2^2$$

Case 2. The other extreme case is when V and W are perpendicular. The slope of the line containing V is $\frac{v_2}{v_1}$. Similarly, the slope of the line containing W is $\frac{w_2}{w_1}$. If these lines are perpendicular then their slopes are negative reciprocals, so $\frac{v_2}{v_1} = -\frac{w_1}{w_2}$. Cross multiplying and bringing everything to one side of the equation then gives $v_1 w_1 + v_2 w_2 = 0$. But the quantity on the left is precisely the definition of $V \cdot W$. To verify the cosine formula just note that if V and W are perpendicular then $\cos \theta = 0$, so $|V||W| \cos \theta = 0$ as well.

EXAMPLE 7-6
The cosine of the angle between vectors $\langle 2, 3 \rangle$ and $\langle -1, 2 \rangle$ can be found using the dot product

$$\langle 2, 3 \rangle \cdot \langle -1, 2 \rangle = -2 + 6 = 4$$
$$= |\langle 2, 3 \rangle||\langle -1, 2 \rangle| \cos \theta$$
$$= \sqrt{4 + 9}\sqrt{1 + 4} \cos \theta$$
$$= \sqrt{65} \cos \theta$$

Hence,

$$\cos \theta = \frac{4}{\sqrt{65}} = \frac{4\sqrt{65}}{65}$$

EXAMPLE 7-7
Since the cosine of the angle between two vectors is zero if and only if the vectors are perpendicular, the dot product can be used to detect this. For example,

$$\langle 2, 3 \rangle \cdot \langle -6, 4 \rangle = -12 + 12 = 0$$

so these vectors must be perpendicular.

Problem 61 *Compute the dot product of the following pairs of vectors.*

1. $V = \langle 2, 4 \rangle$, $W = \langle -3, 1 \rangle$
2. $V = \langle 0, 7 \rangle$, $W = \langle 5, 2 \rangle$
3. $V = \langle -2, -1 \rangle$, $W = \langle 6, -3 \rangle$

Problem 62 *Compute the cosine of the angle between the following pairs of vectors.*

1. $V = \langle 2, 4 \rangle$, $W = \langle -3, 1 \rangle$
2. $V = \langle -2, -1 \rangle$, $W = \langle 6, -3 \rangle$

Problem 63 *Suppose V and W are vectors of length 3. What are the smallest and largest possible values for V · W? What is the smallest possible value of |V · W|?*

Problem 64 *Use the dot product to decide if each of the following pairs of vectors are perpendicular.*

1. $V = \langle 2, 4 \rangle$, $W = \langle -4, 8 \rangle$
2. $V = \langle 2, 4 \rangle$, $W = \langle 8, -4 \rangle$
3. $V = \langle -2, -1 \rangle$, $W = \langle -3, 6 \rangle$
4. $V = \langle -2, -1 \rangle$, $W = \langle 6, -3 \rangle$

7.3 Gradient Vectors and Directional Derivatives

In this section we return to the concept of differentiation of functions of two variables. Recall that the partial derivative $\frac{\partial f}{\partial x}$ was the rate of change of the function $f(x, y)$ if you walk in the positive x-direction with unit speed. Similarly, the partial derivative $\frac{\partial f}{\partial y}$ is the rate of change if you walk in the positive y-direction with unit speed. We are left with the obvious question, "What is the rate of change of $f(x, y)$ if you walk in some other direction?"

Let $V = \langle a, b \rangle$ represent a vector which points in the direction of travel. The length of V will be one, to reflect the fact that we are walking with unit speed. We already know two things:

$$\text{Rate of change of } f(x, y) \text{ in direction } \langle 1, 0 \rangle = \frac{\partial f}{\partial x} = 1\frac{\partial f}{\partial x} + 0\frac{\partial f}{\partial y}$$

$$\text{Rate of change of } f(x, y) \text{ in direction } \langle 0, 1 \rangle = \frac{\partial f}{\partial y} = 0\frac{\partial f}{\partial x} + 1\frac{\partial f}{\partial y}$$

It is reasonable, then, to expect

$$\text{Rate of change of } f(x, y) \text{ in direction } \langle a, b \rangle = a\frac{\partial f}{\partial x} + b\frac{\partial f}{\partial y}$$

EXAMPLE 7-8

Suppose $f(x, y) = xy^2$. Then the rate of change of $f(x, y)$, at the point $(2, 1)$, in the direction $\left(\frac{\sqrt{2}}{2}, \frac{\sqrt{2}}{2}\right)$ is

$$\frac{\sqrt{2}}{2}\frac{\partial f}{\partial x}(2, 1) + \frac{\sqrt{2}}{2}\frac{\partial f}{\partial y}(2, 1) = \frac{\sqrt{2}}{2}1 + \frac{\sqrt{2}}{2}4 = \frac{5\sqrt{2}}{2}$$

The rate of change of $f(x, y)$, in the direction of V, is called a *directional derivative*, and is denoted as $\nabla_V f(x, y)$. Hence, we have the formula

$$\nabla_V f(x, y) = a\frac{\partial f}{\partial x} + b\frac{\partial f}{\partial y}$$

But notice that the right side of this equation can be rewritten as a dot product:

$$\nabla_V f(x, y) = \langle a, b \rangle \cdot \left\langle \frac{\partial f}{\partial x}, \frac{\partial f}{\partial y} \right\rangle$$

We can further shorten this equation by coming up with new notation for $\left\langle \frac{\partial f}{\partial x}, \frac{\partial f}{\partial y} \right\rangle$. Henceforth we will call this vector the *gradient* of f, and denote it as ∇f. The equation for $\nabla_V f$ then becomes

$$\nabla_V f = V \cdot \nabla f$$

EXAMPLE 7-9

The gradient of the function $f(x, y) = x^2 y^3$ at the point (x, y) is

$$\nabla f(x, y) = \langle 2xy^3, 3x^2 y^2 \rangle$$

The gradient at the point $(1, 1)$, then, would be $\nabla f(1, 1) = \langle 2, 3 \rangle$.

Recall that the rate of change of a function is also the slope of a tangent line to its graph, as long as you are traveling with speed one. Here's a nice way to think about the situation. Suppose you are climbing a mountain, and you have a good trail map in your hands. Let (x, y) be your coordinates when you locate yourself on the map. The function $f(x, y)$ is your elevation at that point. Now turn your body to face the direction V (on your map). If you sight up or down so that your gaze just grazes the mountainside then you are looking along the tangent line whose slope is given by $\nabla_V f$.

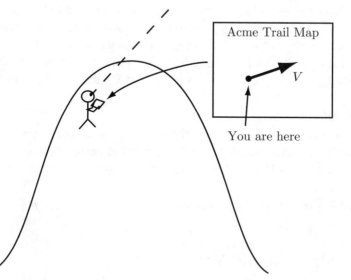

Recall that the dot product is given by the product of the magnitudes of the two vectors, times the cosine of the angle between them. If we fix the point we are at then ∇f is a fixed vector. If $|V| = 1$, then the only way to change $\nabla_V f = V \cdot \nabla f$ is to change the angle between V and ∇f. The largest this quantity can be is when the value of $\cos \theta$ is largest. This happens when $\theta = 0$. We conclude that the largest value of $\nabla_V f$ is given by

$$\nabla_V f = V \cdot \nabla f = |V||\nabla f|\cos \theta = |\nabla f|$$

and that this value is attained when V (the direction we are facing) coincides with the direction of ∇f. In other words, if you are on the mountainside and you want to face directly uphill you should point yourself in the direction of the gradient vector. When you do this and sight along the mountainside the slope you see is the magnitude of the gradient vector.

EXAMPLE 7-10
Let $f(x, y) = e^{xy^2}$. At the point (x, y) the gradient vector is

$$\nabla f(x, y) = \left\langle y^2 e^{xy^2}, 2xy e^{xy^2} \right\rangle$$

So, at the point $(2, 3)$ we have $\nabla f(2, 3) = \langle 9e^{18}, 12e^{18} \rangle$. The largest slope of any tangent line to the graph at the point $(2, 3)$ is then given by

$$|\langle 9e^{18}, 12e^{18} \rangle| = \sqrt{(9e^{18})^2 + (12e^{18})^2} = 15e^{18}$$

What if you were standing on the mountainside and wanted to face the direction you would have to travel to keep your elevation constant? In other words, how would you find the direction of your level curve? If you were facing such a direction you would be looking along a horizontal line, i.e., a line whose slope is zero. The only way for $\nabla_V f = V \cdot \nabla f$ to be zero is for V and ∇f to be perpendicular.

EXAMPLE 7-11
Suppose again $f(x, y) = e^{xy^2}$. In the previous example we saw $\nabla f(2, 3) = \langle 9e^{18}, 12e^{18} \rangle$. A vector which points in a direction perpendicular to this would be $\langle 4, -3 \rangle$ (Check this!). Hence, this vector is tangent to a level curve at the point $(2, 3)$.

Problem 65 *For each of the functions $f(x, y)$ below answer the following questions:*

- *Find the gradient vector $\nabla f(x, y)$.*
- *Calculate the rate of change of $f(x, y)$ at the point $(1, 1)$ in the direction $\left(\frac{3}{5}, \frac{4}{5} \right)$.*
- *Find a unit vector that points in the direction of the maximum rate of change at the point $(1, 1)$.*
- *Find the largest slope of any tangent line at $(1, 1)$.*
- *Find a unit vector that lies in a line tangent to a level curve through $(1, 1)$.*

1. $x \ln y$
2. $2x + 3y$
3. $x^2 y + xy^3$

7.4 Maxima, Minima, and Saddles

At a local maximum or a local minimum of a graph the tangent plane is horizontal. Another way to say this is that the slope of every tangent line to a local maximum or minimum is zero. But recall that the slope of a tangent line at (x, y) in direction V is given by $V \cdot \nabla f$. The only way for this to be zero for every possible V is if ∇f is the zero vector.

EXAMPLE 7-12

We show that $f(x, y) = x^2 y^3$ cannot have a maximum or minimum at any point where x and y are nonzero. The gradient is

$$\nabla f(x, y) = \langle 2xy^3, 3x^2 y^2 \rangle$$

If x and y are nonzero then this is not the zero vector. This tells us that there is a direction where the slopes of tangent lines are nonzero, and hence we cannot be at a local maximum or minimum.

Unfortunately, just because the gradient is the zero vector it does not necessarily mean that there is a local maximum or minimum.

EXAMPLE 7-13

The gradient of $f(x, y) = x^2 + y^2$ and $g(x, y) = x^2 - y^2$ is $\langle 0, 0 \rangle$ at the origin. The function $f(x, y)$ has a minimum there, while $g(x, y)$ has a saddle.

In a first-semester calculus class you learned to detect local maxima and minima by a second-derivative test. We would like to do the same thing here. The problem, of course, is that there are *four* second partial derivatives! To keep track of all this information we often write them in a *matrix*, as follows:

$$\begin{bmatrix} \dfrac{\partial^2 f}{\partial x^2} & \dfrac{\partial^2 f}{\partial x \partial y} \\[2mm] \dfrac{\partial^2 f}{\partial y \partial x} & \dfrac{\partial^2 f}{\partial y^2} \end{bmatrix}$$

Now we examine this matrix for several functions whose graph is familiar. Each of these has a gradient vector equal to zero at the origin.

1. $x^2 + y^2$. This function has a local minimum at the origin. The matrix of second partials is $\begin{bmatrix} 2 & 0 \\ 0 & 2 \end{bmatrix}$.

2. $-x^2 - y^2$. This function has a local maximum at the origin. The matrix of second partials is $\begin{bmatrix} -2 & 0 \\ 0 & -2 \end{bmatrix}$.

3. $x^2 - y^2$. This function has a saddle at the origin. The matrix of second partials is $\begin{bmatrix} 2 & 0 \\ 0 & -2 \end{bmatrix}$.

4. xy. This function also has a saddle at the origin. The matrix of second partials is $\begin{bmatrix} 0 & 1 \\ 1 & 0 \end{bmatrix}$.

The first two give us a clue as to the quantity we would like to look at. Consider the product of the upper-left and lower-right entries of the matrix. For the maximum and minimum above this quantity is positive and for the first of the above saddles it is negative. However, this alone would not be enough to distinguish maxima and minima from saddles, as the second of the saddles shows. To compensate we must subtract the product of the upper-right and lower-left entries, yielding the formula

$$\frac{\partial^2 f}{\partial x^2}\frac{\partial^2 f}{\partial y^2} - \frac{\partial^2 f}{\partial x \partial y}\frac{\partial^2 f}{\partial y \partial x}$$

However, since the mixed partials are equal we can shorten this to

$$\frac{\partial^2 f}{\partial x^2}\frac{\partial^2 f}{\partial y^2} - \left(\frac{\partial^2 f}{\partial x \partial y}\right)^2$$

This is indeed the right quantity to look at, in the sense that if it is greater than zero you have a maximum or minimum, and if it is less than zero you have a saddle. Unfortunately, if it is zero you have no information; you may be at a maximum, minimum, saddle, or something much more bizarre. Nonetheless, we will single this out as our first test.

Test 1

Let $D(x, y) = \frac{\partial^2 f}{\partial x^2}\frac{\partial^2 f}{\partial y^2} - \left(\frac{\partial^2 f}{\partial x \partial y}\right)^2$. Suppose that at some point (x_0, y_0) we have $\nabla f(x_0, y_0) = \langle 0, 0 \rangle$.

$D(x_0, y_0) > 0 \Rightarrow f(x, y)$ has a local max or min at (x_0, y_0).

$D(x_0, y_0) < 0 \Rightarrow f(x, y)$ has a saddle at (x_0, y_0).

$D(x_0, y_0) = 0 \Rightarrow$ No information about $f(x_0, y_0)$.

EXAMPLE 7-14

We find the location of all saddles of the function $f(x, y) = x^2 + 2xy + 3y^3$. First, we will need to narrow down the possibilities by finding the critical points. To do this we find the gradient.

$$\nabla f = \langle 2x + 2y, 2x + 9y^2 \rangle$$

Setting this equal to the vector $\langle 0, 0 \rangle$ gives us the system of equations

$$2x + 2y = 0$$
$$2x + 9y^2 = 0$$

The first equation tells us that $x = -y$. Plugging this into the second equation then gives

$$-2y + 9y^2 = 0$$

Either $y = 0$ (and hence $x = 0$) or we can divide both sides of this equation by y to get

$$-2 + 9y = 0$$

Solving then gives us $y = \frac{2}{9}$. Hence, we have critical points at $(0, 0)$ and $\left(-\frac{2}{9}, \frac{2}{9}\right)$.
   To determine which of these are saddles we compute the matrix of second partials:

$$\begin{bmatrix} \dfrac{\partial^2 f}{\partial x^2} & \dfrac{\partial^2 f}{\partial x \partial y} \\[2ex] \dfrac{\partial^2 f}{\partial y \partial x} & \dfrac{\partial^2 f}{\partial y^2} \end{bmatrix} = \begin{bmatrix} 2 & 2 \\ 2 & 18y \end{bmatrix}$$

And so

$$D(x, y) = 36y - 4$$

We now check each critical point:

$$D(0, 0) = -4 < 0 \Rightarrow (0, 0) \text{ is a saddle.}$$

$$D\left(-\frac{2}{9}, \frac{2}{9}\right) = 4 > 0 \Rightarrow \left(-\frac{2}{9}, \frac{2}{9}\right) \text{ is a local max or min.}$$

   When $D(x_0, y_0) > 0$ it would be nice to have a second test to determine whether (x_0, y_0) is a local maximum or a local minimum. Such a test can be easily guessed from our prototypical examples, $f(x, y) = x^2 + y^2$ and $f(x, y) = -x^2 - y^2$. Notice that in both cases $\frac{\partial^2 f}{\partial x^2}$ and $\frac{\partial^2 f}{\partial y^2}$ have the same sign. When this sign is positive we have a local minimum and when it is negative we have a local maximum. This is precisely our second test.

Test 2

Let $D(x, y) = \frac{\partial^2 f}{\partial x^2} \frac{\partial^2 f}{\partial y^2} - \left(\frac{\partial^2 f}{\partial x \partial y} \right)^2$. Suppose that at some point (x_0, y_0) we have $\nabla f(x_0, y_0) = \langle 0, 0 \rangle$ and $D(x_0, y_0) > 0$.

$$\frac{\partial^2 f}{\partial x^2}(x_0, y_0) > 0 \Rightarrow f(x, y) \text{ has a local min at } (x_0, y_0).$$

$$\frac{\partial^2 f}{\partial x^2}(x_0, y_0) < 0 \Rightarrow f(x, y) \text{ has a local max at } (x_0, y_0).$$

EXAMPLE 7-15
Recall the function $f(x, y) = x^2 + 2xy + 3y^3$ from the previous example. We found critical points at $(0, 0)$ and $\left(-\frac{2}{9}, \frac{2}{9} \right)$, and determined that at $\left(-\frac{2}{9}, \frac{2}{9} \right)$ there was a local maximum or a local minimum. To determine which we need only look at $\frac{\partial^2 f}{\partial x^2}$. Since this was 2, and $2 > 0$, we conclude that at this critical point there is a local minimum.

It is important to keep in mind that if $D(x_0, y_0) = 0$ then we have no information about the nature of $f(x_0, y_0)$. We illustrate this in the next example.

EXAMPLE 7-16
Consider the following functions:

1. $f(x, y) = x^4 + y^4$
2. $f(x, y) = -x^4 - y^4$
3. $f(x, y) = x^2 y^2$

In each case the only critical point is at $(0, 0)$ and $D(0, 0) = 0$. But at $(0, 0)$ in the first case there is a local minimum, in the second there is a local maximum, and in the third there is a saddle.

Problem 66 *Find the local maxima, minima, and saddles of the following functions.*

1. $xy + 2x - 3y - 6$
2. $x^3 - xy + y^2$

Problem 67 *For the function* $\sin(x + y)$ *show that* $D(x, y) = 0$ *at every point* (x, y). *Does this function have maxima, minima, or saddles?*

Problem 68 *If, for some point* (x_0, y_0) *you know* $D(x_0, y_0) > 0$ *and* $\frac{\partial^2 f}{\partial x^2}(x_0, y_0) >$
0 *then show that* $\frac{\partial^2 f}{\partial y^2}(x_0, y_0) > 0$. *(This tells us that you may use* $\frac{\partial^2 f}{\partial y^2}(x_0, y_0)$ *instead*
of $\frac{\partial^2 f}{\partial x^2}(x_0, y_0)$ *when distinguishing maxima from minima.)*

7.5 Application: Optimization Problems

An important application of the mathematics of the previous section is to optimiza-
tion problems. You encountered several such problems in a first semester calculus
class. You can now handle much more complicated situations. We illustrate this
with an example.

EXAMPLE 7-17
We find the largest volume of a box with an open top, and surface area 100 m².
First, we let the dimensions of the box be a, b, and c, with c the height. Then the
surface area is given by

$$S.A. = 100 = ab + 2ac + 2bc$$

Solving this equation for c gives

$$c = \frac{100 - ab}{2(a + b)}$$

The volume, of course, is the quantity abc. Substituting for c then gives us

$$\text{Volume} = ab\frac{100 - ab}{2(a + b)}$$

Thinking of volume, then, as a function $V(a, b)$ we now search for critical
points. The gradient is a bit tedious to compute, but with some work you can show
it simplifies to

$$\nabla V(a, b) = \left\langle \frac{-2a^2b^2 + 200b^2 - 4ab^3}{4(a + b)^2}, \frac{-2a^2b^2 + 200a^2 - 4a^3b}{4(a + b)^2} \right\rangle$$

Each component of this vector can only be zero if the numerators are zero, so
we get two equations:

$$0 = -2a^2b^2 + 200b^2 - 4ab^3$$
$$0 = -2a^2b^2 + 200a^2 - 4a^3b$$

We may safely assume that neither a nor b are zero. Dividing the first equation by $2b^2$ and the second by $2a^2$ then gives

$$0 = -a^2 + 100 - 2ab \qquad (7\text{-}1)$$
$$0 = -b^2 + 100 - 2ab$$

Solving both equations for $2ab$ and setting them equal to each other gives us

$$-a^2 + 100 = -b^2 + 100$$

and thus (since we are only interested in positive values of a and b) we may conclude $a = b$. Combining this with Equation 7-1 tells us

$$0 = -a^2 + 100 - 2a^2$$

Solving this for a reveals

$$a = \frac{10\sqrt{3}}{3}$$

The fact that $a = b$ then gives us

$$b = \frac{10\sqrt{3}}{3}$$

Finally, our expression for c in terms of a and b above yields

$$c = \frac{5\sqrt{3}}{3}$$

The volume is thus

$$\text{Volume} = abc = \frac{500\sqrt{3}}{9}$$

7.6 LaGrange Multipliers

Suppose you wanted to look for the maximum value of $f(x) = \frac{1}{3}x^3 - x$, on the interval $[-2, 3]$. This function has two critical points, and the one at $x = -1$ is a local maximum. But to conclude that $f(-1) = \frac{2}{3}$ is the maximum value of $f(x)$ on the indicated domain would be incorrect. This is because the maximum is attained at one of the endpoints of the domain, namely at $x = 3$. At this point $f(3) = 6$.

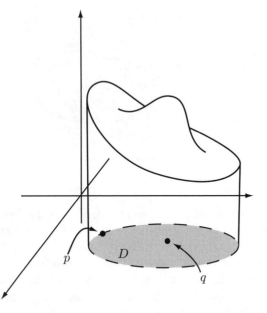

Figure 7-2   The function $f(x, y)$ has a local maximum at the point q, but the
absolute maximum on the domain D is the point p on the boundary

The same kind of thing can happen with functions $f(x, y)$ of two variables. Using techniques of the previous section we can look for, and analyze, critical points in the interior of the domain D of f. But if we are interested in the maximum value that f attains on D we may have to look at the points around the boundary of D, as in Figure 7-2. In the one-variable case this was easy, since it was just a matter of checking two points. When there are multiple variables there are an infinite number of points around the boundary of D, so we need some technique other than just plugging values into f.

In Figure 7-3 we see the contours for the function $f(x, y)$ of Figure 7-2, super-imposed on the domain D. As is customary, we will denote the points around the boundary of D as ∂D. (Don't get this confused with a partial derivative.) Notice that the maximum and minimum values of f on ∂D occur at the places where the level curves of f are tangent to ∂D.

The key fact that we will use is that the vectors ∇f are always perpendicular to the level curves of f. So, at a point $p = (x_0, y_0)$ where a level curve of f is tangent to ∂D we will see $\nabla f(x_0, y_0)$ perpendicular to ∂D.

The next step is to find a new function, $g(x, y)$, for which ∂D is a level curve. The vector ∇g is always perpendicular to all of its level curves. In particular, $\nabla g(x_0, y_0)$ will be perpendicular to ∂D at (x_0, y_0). See Figure 7-4.

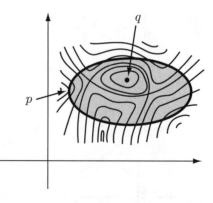

Figure 7-3 The contours of $f(x, y)$ and the domain D. The absolute maximum of f on D is at the point p where a level curve is tangent to ∂D

Since $\nabla f(x_0, y_0)$ and $\nabla g(x_0, y_0)$ are perpendicular to the same curve they must be parallel to each other. If vectors V and W are parallel, then one is a scalar multiple of the other. In other words, there is a constant λ such that $V = \lambda W$. Putting all this together, we find that if a maximum (or minimum) of f occurs at a point (x_0, y_0) of ∂D then there must be a constant λ such that $\nabla f(x_0, y_0) = \lambda \nabla g(x_0, y_0)$. Using this information we get a finite number of candidate points at which to look for maxima and minima.

EXAMPLE 7-18

Let $f(x, y) = 2x^2 + 3y^3$, and suppose D is the set of points (x, y) in $\mathbb{R}^2$ satisfying $x^2 + y^2 \leq 1$. We will look for the largest value attained by $f(x, y)$ on ∂D.

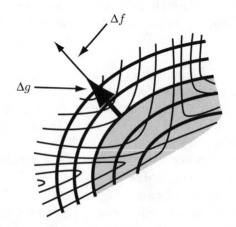

Figure 7-4 The vector ∇f is a scalar multiple of the vector ∇g

To use the "method of Lagrange multipliers" the first step is to find a function $g(x, y)$ for which ∂D is a level curve. This is very easy: $g(x, y) = x^2 + y^2$. Then the level curve $g(x, y) = 1$ is the curve $x^2 + y^2 = 1$, which is precisely the boundary of D.

The next step is to look for points (x_0, y_0), and values of λ, such that $\nabla f(x_0, y_0) = \lambda \nabla g(x_0, y_0)$. We thus compute

$$\nabla f = \langle 4x, 9y^2 \rangle$$

$$\nabla g = \langle 2x, 2y \rangle$$

So if $\langle 4x, 9y^2 \rangle = \lambda \langle 2x, 2y \rangle$ we have $4x = \lambda 2x$ and $9y^2 = \lambda 2y$. The first of these equations tells us that either $x = 0$ or $\lambda = 2$.

Case 1. $x = 0$. The boundary of D is the set of points satisfying $x^2 + y^2 = 1$. So if $x = 0$ then $y = \pm 1$. This means that we must check the points $(0, 1)$ and $(0, -1)$ for maxima and minima.

Case 2. $\lambda = 2$. Combining this with the equations $9y^2 = \lambda 2y$ then tells us $9y^2 = 4y$. So again there are two cases: either $y = 0$ or $9y = 4$, in which case $y = \frac{4}{9}$.

Subcase A. $y = 0$. Similar to Case 1 above, the boundary of D is the points satisfying $x^2 + y^2 = 1$. If $y = 0$ then $x = \pm 1$. So we get two more points to check: $(1, 0)$ and $(-1, 0)$.

Subcase B. $y = \frac{4}{9}$. Plugging this into $x^2 + y^2 = 1$ gives us $x^2 + \frac{16}{81} = 1$. Solving for x then gives $x = \pm \frac{\sqrt{64}}{9}$. Thus we get two more points: $(\frac{\sqrt{64}}{9}, \frac{4}{9})$ and $(-\frac{\sqrt{64}}{9}, \frac{4}{9})$.

We now have six points at which to look for maxima and minima of f. We do this simply by plugging them in:

$$f(0, 1) = 3 \qquad\qquad f(0, -1) = -3$$

$$f(\pm 1, 0) = 2 \qquad\qquad f\left(\pm \frac{\sqrt{64}}{9}, \frac{4}{9} \right) = \frac{454}{243}$$

So the maximum value of f on ∂D is at $(0, 1)$ and the minimum value is at $(0, -1)$.

Problem 69 *Let* $f(x, y) = x^2 + y^2 + 2y - 1$ *and* $D = \{(x, y) | x^2 + \frac{y^2}{4} \leq 1\}$.

  1. *Find the locations of all critical points of* $f(x, y)$ *on D.*
  2. *Find the absolute minimum of* $f(x, y)$ *on D.*

7.7 Determinants

In the previous chapter we saw that the dot product was a way to take two vectors and produce a meaningful number. In two dimensions there is another such way. Suppose $V = \langle a, b \rangle$ and $W = \langle c, d \rangle$ are vectors and consider the parallelogram formed by two parallel copies of each.

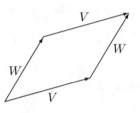

We wish to determine the area of this parallelogram. Notice in the figure below that there are four triangles and two small rectangles inside a larger rectangle. The area of the desired parallelogram is found by subtracting these areas.

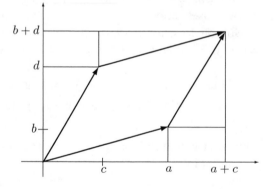

The area of the large parallelogram is $(a + c)(b + d)$. The area of each small rectangle is bc. The area of two of the triangles is $\frac{1}{2}ab$ and the other two is $\frac{1}{2}cd$. Thus the desired area is

$$\text{Area} = (a + c)(b + d) - 2bc - 2\left(\frac{1}{2}ab\right) - 2\left(\frac{1}{2}cd\right)$$

$$= ab + ad + bc + cd - 2bc - ab - cd$$

$$= ad - bc$$

We often organize the information contained in a set of vectors by writing them in a matrix. The area of the parallelogram between them is then the *determinant* of the matrix. We write this as follows:

$$\begin{vmatrix} a & b \\ c & d \end{vmatrix} = ad - bc$$

Notice that there is a bit more information in the determinant than just the area, since our answer may be negative. The sign of the answer tells us the order of V and W. If we go counterclockwise to get from V to W then the sign of the determinant is positive. Otherwise it is negative. For this reason we call the value of the determinant the *signed area of the parallelogram spanned by V and W*.

EXAMPLE 7-19
We compute the (signed) area of the parallelogram spanned by the vectors $\langle 1, 2 \rangle$ and $\langle 4, 3 \rangle$.

$$\begin{vmatrix} 1 & 2 \\ 3 & 4 \end{vmatrix} = 1 \cdot 4 - 2 \cdot 3 = -2$$

In three dimensions there is also a determinant, but it is slightly more complicated. If we take four parallel copies of three vectors then they span a 3-dimensional figure called a *parallelepiped*. You should think of this as a cube that has been stretched, skewed, and rotated. The determinant of the matrix that is comprised of these vectors is then the volume of this figure.

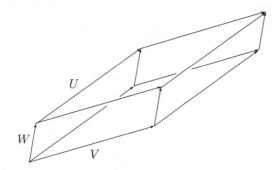

A matrix containing the vectors $U = \langle a, b, c \rangle$, $V = \langle d, e, f \rangle$, and $W = \langle g, h, i \rangle$ would be written like this

$$\begin{bmatrix} a & b & c \\ d & e & f \\ g & h & i \end{bmatrix}$$

To write the determinant of this matrix it is helpful to copy the entire matrix next to the original one. Then add up all the "right-slanting" products and subtract all the "left-slanting" products.

$$\begin{vmatrix} a & b & c \\ d & e & f \\ g & h & i \end{vmatrix}\begin{matrix} a & b & c \\ d & e & f \\ g & h & i \end{matrix} = aei + bfg + cdh - afh - bdi - ceg$$

Once again this answer may be negative, indicating the way in which U, V, and W are situated. The sign is positive if, when we sweep the fingers of our right hand from U to V, our thumb points in the direction of W. Once again, this is the "right-hand rule," that we encountered earlier.

A second way to remember how to compute the determinant of a 3×3 matrix is to reduce it to computing the determinants of three 2×2 matrices:

$$\begin{vmatrix} a & b & c \\ d & e & f \\ g & h & i \end{vmatrix} = a\begin{vmatrix} e & f \\ h & i \end{vmatrix} - b\begin{vmatrix} d & f \\ g & i \end{vmatrix} + c\begin{vmatrix} d & e \\ g & h \end{vmatrix}$$

Notice that each term on the left is preceded by one element from the top row of the matrix, with alternating signs. The 2×2 matrix in each term is the one you get from the original matrix by deleting the corresponding row and column. So, for example, the first term is preceded by a, since that is the first element of the top row. The 2×2 matrix in the first term is the one you get by eliminating the row and column containing a.

EXAMPLE 7-20
We compute the (signed) volume of the parallelepiped spanned by $\langle 1, 2, 3 \rangle$, $\langle 1, -1, 1 \rangle$, and $\langle 0, 2, 2 \rangle$.

$$\begin{vmatrix} 1 & 2 & 3 \\ 1 & -1 & 1 \\ 0 & 2 & 2 \end{vmatrix} = 1(-2 - 2) - 2(2 - 0) + 3(2 - 0) = -4 - 4 + 6 = -2$$

Problem 70 *Find the (signed) area of the parallelogram spanned by*

1. $\langle 1, 3 \rangle$ *and* $\langle -1, 2 \rangle$

2. $\langle 1, 6 \rangle$ *and* $\langle 1, 1 \rangle$

3. $\langle 2, 3 \rangle$ *and* $\langle 6, 9 \rangle$

Problem 71 *Suppose V and W are parallel vectors in* $\mathbb{R}^2$. *Let M be the matrix whose rows are V and W. Show that the determinant of M is zero.*

Problem 72 *Calculate the (signed) area of the parallelepiped spanned by*

1. $\langle 1, 2, 3 \rangle$, $\langle 1, 0, 2 \rangle$, *and* $\langle -2, 2, -3 \rangle$
2. $\langle 0, 1, 3 \rangle$, $\langle -1, 2, 1 \rangle$, *and* $\langle 2, 0, -1 \rangle$

7.8 The Cross Product

A challenging problem is to find the area of a parallelogram defined by two vectors in three dimensions. If the vectors are $V = \langle a, b, c \rangle$ and $W = \langle d, e, f \rangle$ then this area is given by

$$\text{Area}(V, W) = \sqrt{(bf - ec)^2 + (cd - af)^2 + (ae - bd)^2}$$

The general form of this equation looks a lot like the Pythagorean Theorem: some quantity is the square root of the sum of squares. We can employ this observation to make the above formula a bit more compact:

$$\text{Area}(V, W) = |\langle bf - ec, cd - af, ae - bd \rangle|$$

Note that we started with two vectors, V and W, and ended up with a third that was somehow related. This gives us a new kind of product, called the *cross product*:

$$V \times W = \langle bf - ec, cd - af, ae - bd \rangle$$

Recall that the dot product was a way to take two vectors and multiply them, with the result being a scalar. The cross product, by comparison, is a way to multiply and get a third vector.

The astute reader will notice that each component of the cross product is the determinant of a 2×2 matrix. We may thus write its formula like this:

$$V \times W = \left\langle \begin{vmatrix} b & c \\ e & f \end{vmatrix}, \begin{vmatrix} c & a \\ f & d \end{vmatrix}, \begin{vmatrix} a & b \\ d & e \end{vmatrix} \right\rangle$$

It is traditional to define the vectors $\mathbf{i} = \langle 1, 0, 0 \rangle$, $\mathbf{j} = \langle 0, 1, 0 \rangle$, and $\mathbf{k} = \langle 0, 0, 1 \rangle$. Using this we may rewrite the definition of the cross product as

$$V \times W = \mathbf{i} \begin{vmatrix} b & c \\ e & f \end{vmatrix} - \mathbf{j} \begin{vmatrix} a & c \\ d & f \end{vmatrix} + \mathbf{k} \begin{vmatrix} a & b \\ d & e \end{vmatrix}$$

We introduced a negative sign in the second term and switched the order of the columns of the second matrix. Now our formula looks just like a determinant:

$$V \times W = \begin{vmatrix} \mathbf{i} & \mathbf{j} & \mathbf{k} \\ a & b & c \\ d & e & f \end{vmatrix}$$

This last formula is probably the easiest way to remember how to compute the cross product of two vectors.

EXAMPLE 7-21
We compute the area of the parallelogram spanned by the vectors $\langle 1, 1, 0 \rangle$ and $\langle 1, 2, 3 \rangle$. To do this we first compute the cross product:

$$\langle 1, 1, 0 \rangle \times \langle 1, 2, 3 \rangle = \begin{vmatrix} \mathbf{i} & \mathbf{j} & \mathbf{k} \\ 1 & 1 & 0 \\ 1 & 2 & 3 \end{vmatrix}$$

$$= \mathbf{i}(3 - 0) - \mathbf{j}(3 - 0) + \mathbf{k}(2 - 1)$$

$$= 3\mathbf{i} - 3\mathbf{j} + \mathbf{k}$$

$$= \langle 3, -3, 1 \rangle$$

The desired area is now the magnitude of this vector:

$$|\langle 3, -3, 1 \rangle| = \sqrt{3^2 + (-3)^2 + 1^2} = \sqrt{19}$$

We have already seen that the magnitude of the cross product has some geometric significance. What about its direction? To get a clue we compute $V \cdot (V \times W)$. Suppose $V = \langle a, b, c \rangle$ and $W = \langle d, e, f \rangle$. Then, we compute:

$$V \cdot (V \times W) = \langle a, b, c \rangle \cdot \langle bf - ec, cd - af, ae - bd \rangle$$

$$= a(bf - ec) + b(cd - af) + c(ae - bd)$$

$$= abf - aec + bcd - baf + cae - cbd$$

$$= 0$$

From this we may conclude that $V \times W$ is perpendicular to V. A similar computation shows $W \cdot (V \times W) = 0$, and hence $V \times W$ is perpendicular to W as well. We conclude $V \times W$ is perpendicular to the *plane* containing V and W.

EXAMPLE 7-22
We find a unit vector that is perpendicular to both $\langle 1, 1, 0 \rangle$ and $\langle 1, 2, 3 \rangle$. From the previous problem we know that the cross product of these two vectors is the vector $\langle 3, -3, 1 \rangle$, and that the magnitude of this vector is $\sqrt{19}$. To get a unit vector which points in the same direction we just divide by the magnitude:

$$\frac{\langle 3, -3, 1 \rangle}{\sqrt{19}} = \left\langle \frac{3\sqrt{19}}{19}, \frac{-3\sqrt{19}}{19}, \frac{\sqrt{19}}{19} \right\rangle$$

Problem 73 *Find the area of the parallelogram spanned by the vectors*

1. $\langle 1, 2, 3 \rangle$ *and* $\langle -1, 0, 1 \rangle$
2. $\langle 1, 1, 0 \rangle$ *and* $\langle 1, 0, 1 \rangle$
3. $\langle 1, 2, 3 \rangle$ *and* $\langle 3, 1, 2 \rangle$

Problem 74 *Find a unit vector which is perpendicular to each of the following pairs of vectors.*

1. $\langle 1, 2, 0 \rangle$ *and* $\langle 1, 1, 1 \rangle$
2. $\langle 1, 1, 2 \rangle$ *and* $\langle 1, 1, 1 \rangle$

Problem 75 *Show that* $V \times W = -W \times V$.

Problem 76 *Let U, V, and W be three vectors in $\mathbb{R}^3$. Let M be the 3×3 matrix whose rows are these vectors. Show that the determinant of M is equal to $U \cdot (V \times W)$.*

Problem 77 *Show that the area of the parallelogram spanned by V and W is equal to $|V||W|\sin\theta$, where θ is the angle between V and W.*

Quiz

Problem 78

1. *Let $f(x, y)$ be the following function:*

$$f(x, y) = x^2 + 3xy$$

    a. *Find the largest slope of any tangent line to the graph of $f(x, y)$ at the point $(1, 1)$.*

    b. *Find the critical point(s).*

    c. *Compute the value of*

$$\begin{vmatrix} \dfrac{\partial^2 f}{\partial x^2} & \dfrac{\partial^2 f}{\partial x \partial y} \\[2ex] \dfrac{\partial^2 f}{\partial y \partial x} & \dfrac{\partial^2 f}{\partial y^2} \end{vmatrix}$$

    for each critical point found.

    d. *Does the graph have a max, min, or saddle at the critical point(s)?*

2. *Let $V = \langle 1, 2, 3 \rangle$ and $W = \langle 1, 1, 1 \rangle$. Find*

    a. *a unit vector which points in the same direction as V.*

    b. *the cosine of the angle between V and W.*

    c. *a unit vector which is perpendicular to both V and W.*

CHAPTER 8

Calculus with Parameterizations

8.1 Differentiating Parameterizations

Suppose $\phi(t) = (f(t), g(t))$ is a parameterization of some curve in $\mathbb{R}^2$. What should it mean to *differentiate* ϕ at $t = t_0$? Obviously, the answer will involve two numbers, $f'(t_0)$ and $g'(t_0)$. The best way to make sense out of this pair of numbers is as a vector:

$$\frac{d}{dt}\phi(t_0) = \langle f'(t_0), g'(t_0) \rangle$$

If this vector is drawn with its base at the point $\phi(t_0)$ then it is tangent to the parameterized curve. In addition, if t really represents time, and $\phi(t)$ the coordinates of a particle, then the *length* of this vector tells how fast the particle is moving at time t_0.

EXAMPLE 8-1

At time t the coordinates of a car are $\phi(t) = (t^2, t^3)$. We wish to determine how fast, and in what direction, the car is heading at time $t = 2$. First, we differentiate the parameterization:

$$\phi'(t) = \langle 2t, 3t^2 \rangle$$

Thus,

$$\phi(2) = \langle 4, 12 \rangle$$

The direction of the car is thus the direction that this vector is pointing. Its speed is given by the magnitude of this vector:

$$|\langle 4, 12 \rangle| = \sqrt{4^2 + 12^2} = 4\sqrt{10}$$

Nothing significant changes in three dimensions, as in our next example.

EXAMPLE 8-2

We find the speed of a particle moving along the helix $\phi(t) = (\cos t, \sin t, t)$. First, we find the tangent vector:

$$\phi'(t) = \langle -\sin t, \cos t, 1 \rangle$$

To find the speed we compute the magnitude of this vector:

$$|\langle -\sin t, \cos t, 1 \rangle| = \sqrt{(-\sin t)^2 + (\cos t)^2 + 1^2} = \sqrt{2}$$

Note that our answer does not depend on t, telling us a particle moving along the spiral by the given parameterization travels with constant speed.

Differentiating parameterizations of surfaces is a bit more complicated, since now there are two variables. For example, consider the general parameterization

$$\Psi(u, v) = (f(u, v), g(u, v), h(u, v))$$

We can take the *partial* derivative of Ψ with respect to u, say, simply by taking the partial derivatives of f, g, and h with respect to u. Our answer is again a vector:

$$\frac{\partial}{\partial u}\Psi = \left\langle \frac{\partial f}{\partial u}, \frac{\partial g}{\partial u}, \frac{\partial h}{\partial u} \right\rangle$$

Geometrically, $\frac{\partial \Psi}{\partial u}(u_0, v_0)$ is a tangent vector to the surface parameterized by Ψ at the point $\Psi(u_0, v_0)$. Another such vector is given by $\frac{\partial \Phi}{\partial v}(u_0, v_0)$.

EXAMPLE 8-3

We find two vectors tangent to the graph of $z = x^2 + y^3$ at the point $(2, 1, 5)$. This surface is parameterized by

$$\Psi(x, y) = (x, y, x^2 + y^3)$$

The desired point is at $\Psi(2, 1)$. To find two tangent vectors we simply take the partial derivatives of Ψ and evaluate at $(2, 1)$.

$$\frac{\partial \Psi}{\partial x} = \langle 1, 0, 2x \rangle$$

and so,

$$\frac{\partial \Psi}{\partial x}(2, 1) = \langle 1, 0, 4 \rangle$$

Similarly,

$$\frac{\partial \Psi}{\partial y} = \langle 0, 1, 3y^2 \rangle$$

and so,

$$\frac{\partial \Psi}{\partial y}(2, 1) = \langle 0, 1, 3 \rangle$$

We conclude $\langle 1, 0, 4 \rangle$ and $\langle 0, 1, 3 \rangle$ are two vectors tangent to the graph of $z = x^2 + y^3$ at the point $(2, 1, 5)$.

EXAMPLE 8-4

If (θ, ϕ) represents your longitude and latitude, then your position on the Earth's surface is given by

$$\Psi(\theta, \phi) = (R \cos \phi \cos \theta, R \cos \phi \sin \theta, R \sin \phi)$$

where R represents the radius of the Earth (assuming the Earth is a perfect sphere). Note the similarity to spherical coordinates. We explore the geometric meaning of the derivatives of this parameterization.

First, we differentiate with respect to ϕ

$$\frac{\partial \Psi}{\partial \phi} = \langle -R \sin \phi \cos \theta, -R \sin \phi \sin \theta, R \cos \phi \rangle$$

The magnitude of this vector is thus

$$\left| \frac{\partial \Psi}{\partial \phi} \right| = \sqrt{(R \sin \phi \cos \theta)^2 + (R \sin \phi \sin \theta)^2 + (R \cos \phi)^2}$$

$$= \sqrt{R^2 \sin^2 \phi + R^2 \cos^2 \phi}$$

$$= R$$

Notice that the result is constant, i.e., it does not depend on your longitude or latitude. The significance of this is that if you change your latitude by roughly 1 radian then you change your position on the Earth by a constant amount. We will see in a moment that this is very different than changing your longitude. Let's try differentiating with respect to θ.

$$\frac{\partial \Psi}{\partial \theta} = \langle -R \cos \phi \sin \theta, R \cos \phi \cos \theta, 0 \rangle$$

The magnitude of this vector is thus

$$\left| \frac{\partial \Psi}{\partial \theta} \right| = \sqrt{(R \cos \phi \sin \theta)^2 + (R \cos \phi \cos \theta)^2} = R \cos \phi$$

This says that if you change your longitude by 1 radian, then the actual change in distance on the Earth's surface is proportional to the cosine of your latitude. Now let's take the cross product of the vectors we have found:

$$\frac{\partial \Psi}{\partial \theta} \times \frac{\partial \Psi}{\partial \phi} = \begin{vmatrix} \mathbf{i} & \mathbf{j} & \mathbf{k} \\ -R \cos \phi \sin \theta & R \cos \phi \cos \theta & 0 \\ -R \sin \phi \cos \theta & -R \sin \phi \sin \theta & R \cos \phi \end{vmatrix}$$

$$= \langle R^2 \cos^2 \phi \cos \theta, R^2 \cos^2 \phi \sin \theta, R^2 \cos \phi \sin \phi \rangle$$

The magnitude of this vector is

$$\left| \frac{\partial \Psi}{\partial \theta} \times \frac{\partial \Psi}{\partial \phi} \right| = \sqrt{(R^2 \cos^2 \phi \cos \theta)^2 + (R^2 \cos^2 \phi \sin \theta)^2 + (R^2 \cos \phi \sin \phi)^2}$$

$$= R^2 \cos \phi$$

This says that if you stake out a parcel of land that is 1 radian of longitude by 1 radian of latitude, then its area is proportional to the cosine of your latitude.

Problem 79 *Find a function $s(t)$ which gives the speed of a particle at time t if its coordinates at time t are given by $(\sin t^2, \cos t^2)$.*

Problem 80 *A curve in $\mathbb{R}^3$ is parameterized by $c(t) = (t^3, t^4, t^5)$. Find a unit vector which is tangent to this curve at the point $(1, 1, 1)$.*

Problem 81 *Let S be the paraboloid parameterized by*

$$\Psi(r, \theta) = (r \cos \theta, r \sin \theta, r^2)$$
$$0 \le r \le 2, \quad 0 \le \theta \le 2\pi$$

1. *Find two (nonparallel) vectors tangent to S at the point $\Psi\left(1, \frac{\pi}{6}\right)$.*
2. *Find a vector which is perpendicular to S at the point $\Psi\left(1, \frac{\pi}{6}\right)$. (Hint: Use the cross product.)*

Problem 82 *Let $\Psi(t) = (\cos t, \sin t, t)$.*

1. *Find a parameterization Φ of the curve parameterized by Ψ in which $\left|\frac{d\Phi}{dt}\right| = 1$.*
2. *We may think of $\frac{d\Phi}{dt}$ as itself a parameterized curve. For every value of t we picture the vector $\frac{d\Phi}{dt}$ as being based at the origin, and pointing to some point in $\mathbb{R}^3$. We may thus differentiate this parameterization to obtain another vector $\frac{d^2\Phi}{dt^2}$. Show that $\frac{d\Phi}{dt}$ is perpendicular to $\frac{d^2\Phi}{dt^2}$. (Hint: Compute $\frac{d^2\Phi}{dt^2}$ and show that the dot product with $\frac{d\Phi}{dt}$ is zero.)*
3. *Find κ so that $\frac{d^2\Phi}{dt^2} = \kappa N$, where N is a unit vector. (The number κ is called the curvature of the parameterized curve and N is called the normal.)*
4. *Compute $B = \frac{d\Phi}{dt} \times N$ and show $|B| = 1$. (The vector B is called the binormal.)*
5. *Compute $\frac{dB}{dt}$.*
6. *Find τ such that $\frac{dB}{dt} = -\tau N$. (The number τ is called the torsion.)*
7. *Show that*

$$\frac{dN}{dt} = -\kappa \frac{d\Phi}{dt} + \tau B$$

8.2 Arc Length

Suppose C is a curve, in $\mathbb{R}^2$ or $\mathbb{R}^3$, which is parameterized by the function $\Psi(t)$, where $a \leq t \leq b$. We would like to find the length of C. We do this by sampling C at various points, looking at the length of the line segment connecting adjacent points, and adding these lengths. The result will be an approximation of the length of C. This approximation will get better and better as we sample more and more points along C. In the resulting limit, we obtain an integral which represents the length.

The steps are very straightforward.

1. Choose n points in the interval $[a, b]$. We will call these points $\{t_i\}_{i=0}^{n}$.

2. Note that $\Psi(t_i)$ is a point of C. For each i we connect $\Psi(t_{i+1})$ to $\Psi(t_i)$ by a line segment to approximate C.

3. The length of each line segment is precisely the length of the vector $\Psi(t_{i+1}) - \Psi(t_i)$.

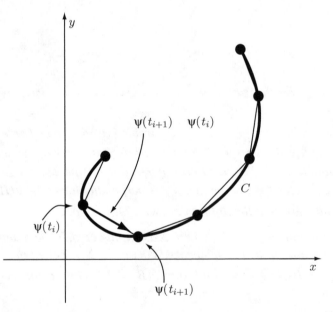

4. We now add up the lengths of all the approximating line segments to get an approximation for the length of C:

$$\sum_{i=0}^{n} |\Psi(t_{i+1}) - \Psi(t_i)|$$

A little algebraic trick will be useful in the next step:

$$\sum_{i=0}^{n} |\Psi(t_{i+1}) - \Psi(t_i)| = \sum_{i=0}^{n} |\Psi(t_{i+1}) - \Psi(t_i)| \frac{\Delta t}{\Delta t}$$

$$= \sum_{i=0}^{n} \left| \frac{\Psi(t_{i+1}) - \Psi(t_i)}{\Delta t} \right| \Delta t$$

5. Finally, we take the limit of this quantity as $n \to \infty$. Note that we can also view this as a limit as $\Delta t \to 0$. But, by definition,

$$\lim_{\Delta t \to 0} \frac{\Psi(t_{i+1}) - \Psi(t_i)}{\Delta t} = \frac{d\Psi}{dt}$$

Hence, the desired limit is

$$\lim_{n \to \infty} \sum_{i=0}^{n} \left| \frac{\Psi(t_{i+1}) - \Psi(t_i)}{\Delta t} \right| \Delta t = \int_a^b \left| \frac{d\Psi}{dt} \right| dt$$

EXAMPLE 8-5

Let C be the curve parameterized by

$$\Psi(t) = (t \cos t, t \sin t), \quad 0 \le t \le 1$$

Then the length of C is given by

$$\int_0^1 \left| \frac{d\Psi}{dt} \right| dt = \int_0^1 |\langle \cos t - t \sin t, \sin t + t \cos t \rangle| \, dt$$

$$= \int_0^1 \sqrt{(\cos t - t \sin t)^2 + (\sin t + t \cos t)^2} \, dt$$

$$= \int_0^1 \sqrt{1 + t^2} \, dt$$

$$= \frac{1}{2} t \sqrt{1 + t^2} + \frac{1}{2} \ln \left| t + \sqrt{1 + t^2} \right| \Big|_0^1$$

$$= \frac{\sqrt{2}}{2} + \frac{1}{2} \ln(1 + \sqrt{2})$$

Problem 83 *Calculate the length of the curve in $\mathbb{R}^3$ parameterized by*

$$\Psi(t) = (\cos t, \sin t, t)$$

$$0 \leq t \leq 1$$

Problem 84 *Show that the curve in $\mathbb{R}^3$ parameterized by*

$$\Psi(t) = \left(\cos t, \sin t, \frac{1}{2}t^2 \right)$$

$$0 \leq t \leq 1$$

has the same length as the curve in $\mathbb{R}^2$ parameterized by

$$(t \cos t, t \sin t), \quad 0 \leq t \leq 1$$

8.3 Line Integrals

Let C represent some parameterized curve in $\mathbb{R}^2$. Let $f(x, y)$ be a function on $\mathbb{R}^2$. In this section we make some sense of the integral of $f(x, y)$ over C. First, let's take a step backward and recall the steps used to define the integral of a function of one variable, $f(x)$, over an interval $[a, b]$.

1. Choose n points in $[a, b]$, which we denote as $\{x_i\}$.
2. Let $\Delta x_i = x_{i+1} - x_i$.
3. For each i compute $f(x_i)\Delta x_i$.
4. Sum over all i.
5. Define $\int_a^b f(x)\,dx$ to be the limit of the sum from the previous step as $n \to \infty$.

The trick to defining the integral of $f(x, y)$ over a parameterized curve C is to follow the above steps as closely as possible. As we will see presently, the most difficult step is probably the one that seems most innocent—the second.

1. Choose n points along the curve C, which we denote as $\{p_i\}$. (Note that p_i is a label for a point of $\mathbb{R}^2$, and as such will have two coordinates.)
2. Let δp_i denote the length, along C, from p_i to p_{i+1}.
3. For each i compute $f(p_i)\delta p_i$.
4. Sum over all i.
5. Define $\int_C f(x, y)\,ds$ to be the limit of the sum from the previous step as $n \to \infty$.

The resulting integral is called the *line integral* of $f(x, y)$. Unfortunately, in practice the above steps are impossible to carry out. To make things easier we employ a parameterization of C to translate the problem to an integral of a function of one variable over an interval of $\mathbb{R}^1$. To this end, let $\Psi(t)$, where $a \leq t \leq b$, be a parameterization of C. We now repeat the above steps as closely as possible, this time using Ψ.

1. Choose n points in the interval $[a, b]$. We denote these points as $\{t_i\}$. The set $\{\Psi(t_i)\}$ is then a collection of n points on C.

2. Let $\Delta t_i = t_{i+1} - t_i$. Let δt_i denote the length, along C, from $\{\Psi(t_i)\}$ to $\{\Psi(t_{i+1})\}$.

3. For each i compute $f(\Psi(t_i))\delta t_i$. A little algebraic trickery yields

$$f(\Psi(t_i))\delta t_i = f(\Psi(t_i))\delta t_i \frac{\Delta t_i}{\Delta t_i}$$

$$= f(\Psi(t_i))\frac{\delta t_i}{\Delta t_i}\Delta t_i$$

4. Sum over all i.

5. The integral $\int_C f(x, y)\, d\mathbf{s}$ is defined to be the limit of the sum from the previous step as $n \to \infty$. But as $n \to \infty$ it also follows that $\Delta t_i \to 0$. Our integrand contains the term $\frac{\delta t_i}{\Delta t_i}$. As $\Delta t_i \to \infty$ this converges to $\left|\frac{\partial \Psi}{\partial t}\right|$. Hence, our integral has become

$$\int_C f(x, y)\, d\mathbf{s} = \int_a^b f(\Psi(t))\left|\frac{\partial \Psi}{\partial t}\right| dt$$

Note that if $f(x, y) = 1$ then the line integral of f over C gives precisely the length of C. This gives us one way to think about line integrals. Imagine a vast plane. At every point of the plane there is a number which represents how fast a point moving through that point will travel. If this number is always one, then a point moving along in the plane will always travel with unit speed. Its total travel time will then be exactly the same as the total distance it travels. But if the number at each point isn't always one then its travel time may be different than the total distance traveled. The line integral gives the total travel time.

EXAMPLE 8-6
We integrate the function $f(x, y) = y$ over the top half of the circle of radius 1. This curve is parameterized by

$$\Psi(t) = (\cos t, \sin t), \quad 0 \leq t \leq \pi$$

We now compute

$$\int_C f(x, y)\, d\mathbf{s} = \int_0^\pi f(\cos t, \sin t)\, |\langle -\sin t, \cos t\rangle|\ dt$$

$$= \int_0^\pi \sin t\, |\langle -\sin t, \cos t\rangle|\ dt$$

$$= \int_0^\pi \sin t \sqrt{\sin^2 t + \cos^2 t}\ dt$$

$$= \int_0^\pi \sin t\ dt$$

$$= -\cos t\big|_0^\pi$$

$$= 2$$

Problem 85 *Let $f(x, y) = y^3$. Compute the line integral of $f(x, y)$ over the curve C parameterized by*

$$\Psi(t) = \left(\frac{1}{3}t^3, t\right), \quad 0 \le t \le 1$$

8.4 Surface Area

One of the most interesting applications of multiple integrals is the computation of surface area. Let R be a rectangle in the xy-plane. The problem is to find the area of the portion of the surface parameterized by $\Psi(u, v)$. Let R denote the domain of Ψ. To find the area we follow similar steps that we used to derive the formula for the volume under $f(x, y)$.

1. We begin by choosing a grid of points $\{(u_i, v_j)\}$ in R.
2. For each grid point locate the point $\Psi(u_i, v_j)$ in $\mathbb{R}^3$. Connecting adjacent points produces a bunch of parallelograms that piece together to form an approximation of the parameterized surface.

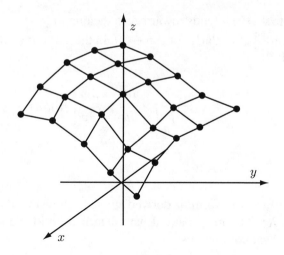

3. We now compute the area of each parallelogram. Observe that each parallelogram is spanned by the vectors

$$V_u = \Psi(u_{i+1}, v_j) - \Psi(u_i, v_j)$$

and

$$V_v = \Psi(u_i, v_{j+1}) - \Psi(u_i, v_j)$$

The desired area is thus the magnitude of the cross product of these vectors:

$$\text{Area} = |V_u \times V_v|$$

We now do some algebraic tricks:

$$|V_u \times V_v| = |V_u \times V_v| \frac{\Delta u \Delta v}{\Delta u \Delta v}$$

$$= \left| \frac{V_u}{\Delta u} \times \frac{V_v}{\Delta v} \right| \Delta u \Delta v$$

4. The surface area is the limit, as Δu and Δv tend toward zero, of the sum of this quantity over all i and j:

$$S.A. = \lim_{\Delta u, \Delta v \to 0} \sum_{i,j} \left| \frac{V_u}{\Delta u} \times \frac{V_v}{\Delta v} \right| \Delta u \Delta v$$

But notice that as Δu tends toward zero the quantity $\frac{V_u}{\Delta u} = \frac{\Psi(u_{i+1}, v_j) - \Psi(u_i, v_j)}{\Delta u}$ converges on $\frac{\partial \Psi}{\partial u}$. Similarly for $\Delta v \to 0$. In the end our summation becomes the integral:

$$S.A. = \iint_R \left| \frac{\partial \Psi}{\partial u} \times \frac{\partial \Psi}{\partial v} \right| \, du \, dv$$

EXAMPLE 8-7

We will use the surface area formula derived above to show that the area of a sphere of radius R is $4\pi R^2$. The most natural parameterization of the sphere comes, of course, from spherical coordinates:

$$\Psi(\theta, \phi) = (R \sin \phi \cos \theta, \, R \sin \phi \sin \theta, \, R \cos \phi)$$

$$0 \le \theta \le 2\pi, \quad 0 \le \phi \le \pi$$

The partial derivatives of this parameterization are

$$\frac{\partial \Psi}{\partial \theta} = \langle -R \sin \phi \sin \theta, \, R \sin \phi \cos \theta, \, 0 \rangle$$

$$\frac{\partial \Psi}{\partial \phi} = \langle R \cos \phi \cos \theta, \, R \cos \phi \sin \theta, \, -R \sin \phi \rangle$$

The cross product of these two vectors is thus

$$\frac{\partial \Psi}{\partial \theta} \times \frac{\partial \Psi}{\partial \phi} = \begin{vmatrix} \mathbf{i} & \mathbf{j} & \mathbf{k} \\ -R \sin \phi \sin \theta & R \sin \phi \cos \theta & 0 \\ R \cos \phi \cos \theta & R \cos \phi \sin \theta & -R \sin \phi \end{vmatrix}$$

$$= -R^2 \sin^2 \phi \cos \theta \mathbf{i} - R^2 \sin^2 \phi \sin \theta \mathbf{j}$$

$$- (R^2 \sin \phi \cos \phi \sin^2 \theta + R^2 \sin \phi \cos \phi \cos^2 \theta) \mathbf{k}$$

$$= -R^2 \sin^2 \phi \cos \theta \mathbf{i} - R^2 \sin^2 \phi \sin \theta \mathbf{j} - R^2 \sin \phi \cos \phi \mathbf{k}$$

The magnitude of this vector is

$$\left| \frac{\partial \Psi}{\partial \theta} \times \frac{\partial \Psi}{\partial \phi} \right|$$

$$= \sqrt{(-R^2 \sin^2 \phi \cos \theta)^2 + (-R^2 \sin^2 \phi \sin \theta)^2 + (-R^2 \sin \phi \cos \phi)^2}$$

$$= \sqrt{R^4 \sin^4 \phi \cos^2 \theta + R^4 \sin^4 \phi \sin^2 \theta + R^4 \sin^2 \phi \cos^2 \phi}$$

$$= R^2 \sqrt{\sin^4 \phi + \sin^2 \phi \cos^2 \phi}$$

$$= R^2 \sin \phi \sqrt{\sin^2 \phi + \cos^2 \phi}$$

$$= R^2 \sin \phi$$

The desired surface area is thus computed as follows:

$$S.A. = \int_0^\pi \int_0^{2\pi} \left| \frac{\partial \Psi}{\partial \theta} \times \frac{\partial \Psi}{\partial \phi} \right| \, d\theta d\phi$$

$$= \int_0^\pi \int_0^{2\pi} R^2 \sin \phi \, d\theta d\phi$$

$$= 2\pi R^2 \int_0^\pi \sin \phi \, d\phi$$

$$= 2\pi R^2 \left(-\cos \phi \right)\big|_0^\pi$$

$$= 2\pi R^2 (1 + 1)$$

$$= 4\pi R^2$$

There are lots of special cases where one may want to compute surface area. In each of these cases we can derive an appropriate formula. The first case we consider is the surface area of the portion of the graph of $z = f(x, y)$ that lies above some domain R. Such a graph can be parameterized as follows:

$$\Psi(x, y) = (x, y, f(x, y))$$

The partial derivatives of this parameterization are

$$\frac{\partial \Psi}{\partial x} = \left\langle 1, 0, \frac{\partial f}{\partial x} \right\rangle$$

$$\frac{\partial \Psi}{\partial y} = \left\langle 0, 1, \frac{\partial f}{\partial y} \right\rangle$$

The cross product of these vectors is

$$\frac{\partial \Psi}{\partial x} \times \frac{\partial \Psi}{\partial y} = \begin{vmatrix} \mathbf{i} & \mathbf{j} & \mathbf{k} \\ 1 & 0 & \frac{\partial f}{\partial x} \\ 0 & 1 & \frac{\partial f}{\partial y} \end{vmatrix}$$

$$= -\frac{\partial f}{\partial x}\mathbf{i} - \frac{\partial f}{\partial y}\mathbf{j} + \mathbf{k}$$

The magnitude of this vector is

$$\left| \frac{\partial \Psi}{\partial x} \times \frac{\partial \Psi}{\partial y} \right| = \sqrt{\left(\frac{\partial f}{\partial x}\right)^2 + \left(\frac{\partial f}{\partial y}\right)^2 + 1}$$

The desired formula for the surface area of a graph is thus

$$S.A. = \iint_R \left| \frac{\partial \Psi}{\partial x} \times \frac{\partial \Psi}{\partial y} \right| \, dx \, dy$$

$$= \iint_R \sqrt{\left(\frac{\partial f}{\partial x}\right)^2 + \left(\frac{\partial f}{\partial y}\right)^2 + 1} \, dx \, dy$$

EXAMPLE 8-8
We will use the surface area formula derived above for the graph of a function to once again show that the area of a sphere of radius r is $4\pi r^2$. An equation for the top half of the sphere is $z = \sqrt{r^2 - x^2 - y^2}$. We will compute the area of this surface and double it to get the total area of the sphere.

The tricky part of this computation is the limits of integration. Notice that the domain of integration is the region inside a circle of radius r in the xy-plane. For a fixed x the value of y can range from $-\sqrt{r^2 - x^2}$ to $\sqrt{r^2 - x^2}$. Hence, we may set up a double integral to compute surface area as follows:

$$\int\limits_{-r}^{r} \int\limits_{-\sqrt{r^2-x^2}}^{\sqrt{r^2-x^2}} \sqrt{\left(\frac{\partial f}{\partial x}\right)^2 + \left(\frac{\partial f}{\partial y}\right)^2 + 1}\, dy\, dx$$

$$= \int\limits_{-r}^{r} \int\limits_{-\sqrt{r^2-x^2}}^{\sqrt{r^2-x^2}} \sqrt{\left(\frac{-x}{\sqrt{r^2-x^2-y^2}}\right)^2 + \left(\frac{-y}{\sqrt{r^2-x^2-y^2}}\right)^2 + 1}\, dy\, dx$$

$$= \int\limits_{-r}^{r} \int\limits_{-\sqrt{r^2-x^2}}^{\sqrt{r^2-x^2}} \sqrt{\frac{x^2 + y^2 + r^2 - x^2 - y^2}{r^2 - x^2 - y^2}}\, dy\, dx$$

$$= \int\limits_{-r}^{r} \int\limits_{-\sqrt{r^2-x^2}}^{\sqrt{r^2-x^2}} \frac{r}{\sqrt{r^2 - x^2 - y^2}}\, dy\, dx$$

$$= \int\limits_{-r}^{r} r \sin^{-1}\left(\frac{y}{\sqrt{r^2 - x^2}}\right)\Bigg|_{-\sqrt{r^2-x^2}}^{\sqrt{r^2-x^2}}\, dx$$

$$= r \int\limits_{-r}^{r} \sin^{-1}(1) - \sin^{-1}(-1)\, dx$$

$$= r \int\limits_{-r}^{r} \pi\, dx$$

$$= \pi r x \big|_{-r}^{r}$$

$$= 2\pi r^2$$

As this is the area of just the top half of the sphere we double it to obtain $4\pi r^2$, the correct formula.

Another special case of a surface area formula is for a surface of revolution. This formula is typically covered in a second semester calculus class. We will show here how this formula can be derived from the one above.

Suppose we begin with the graph of a function $z = f(x)$, where $0 \le a \le x \le b$, and get a surface by revolving it around the z-axis. The resulting surface is best parameterized with cylindrical coordinates, since it has the equation $z = f(r)$.

$$\Psi(r, \theta) = (r \cos \theta, r \sin \theta, f(r))$$

$$a \le r \le b, \quad 0 \le \theta \le 2\pi$$

The partial derivatives are

$$\frac{\partial \Psi}{\partial r} = \left\langle \cos \theta, \sin \theta, \frac{df}{dr} \right\rangle$$

$$\frac{\partial \Psi}{\partial \theta} = \langle -r \sin \theta, r \cos \theta, 0 \rangle$$

The cross product of these vectors is

$$\frac{\partial \Psi}{\partial r} \times \frac{\partial \Psi}{\partial \theta} = \begin{vmatrix} \mathbf{i} & \mathbf{j} & \mathbf{k} \\ \cos \theta & \sin \theta & \frac{df}{dr} \\ -r \sin \theta & r \cos \theta & 0 \end{vmatrix}$$

$$= -r \cos \theta \frac{df}{dr} \mathbf{i} - r \sin \theta \frac{df}{dr} \mathbf{j} + (r \cos^2 \theta + r \sin^2 \theta) \mathbf{k}$$

$$= -r \cos \theta \frac{df}{dr} \mathbf{i} - r \sin \theta \frac{df}{dr} \mathbf{j} + r \mathbf{k}$$

The magnitude of this vector is

$$\left| \frac{\partial \Psi}{\partial r} \times \frac{\partial \Psi}{\partial \theta} \right| = \sqrt{\left(-r \cos \theta \frac{df}{dr} \right)^2 + \left(-r \sin \theta \frac{df}{dr} \right)^2 + r^2}$$

$$= \sqrt{r^2 \cos^2 \theta \left(\frac{df}{dr} \right)^2 + r^2 \sin^2 \theta \left(\frac{df}{dr} \right)^2 + r^2}$$

$$= \sqrt{r^2 \left(\frac{df}{dr} \right)^2 + r^2}$$

$$= r \sqrt{\left(\frac{df}{dr} \right)^2 + 1}$$

The desired surface area is now computed as an integral:

$$S.A. = \int_a^b \int_0^{2\pi} \left| \frac{\partial \Psi}{\partial r} \times \frac{\partial \Psi}{\partial \theta} \right| \ d\theta \ dr$$

$$= \int_a^b \int_0^{2\pi} r \sqrt{\left(\frac{df}{dr} \right)^2 + 1} \ d\theta \ dr$$

$$= 2\pi \int_a^b r \sqrt{\left(\frac{df}{dr} \right)^2 + 1} \ dr$$

EXAMPLE 8-9

We compute the surface area of a sphere one final time. This time we think of the sphere as a surface of revolution and use the formula just derived. As in the previous example we will compute the area of the top half and double it.

To think of the top half of the sphere as a surface of revolution we begin with a function which gives part of a circle:

$$f(x) = \sqrt{R^2 - x^2}, \quad 0 \le x \le R$$

Applying the above formula for the surface area of a surface of revolution gives the following integral:

$$S.A. = 2\pi \int_0^R x \sqrt{\left(\frac{-x}{\sqrt{R^2 - x^2}} \right)^2 + 1} \ dx$$

$$= 2\pi \int_0^R x \sqrt{\frac{x^2}{R^2 - x^2} + 1} \ dx$$

$$= 2\pi \int_0^R x \sqrt{\frac{R^2}{R^2 - x^2}} \ dx$$

$$= 2\pi \int_0^R \frac{Rx}{\sqrt{R^2 - x^2}}\, dx$$

$$= \pi R \int_0^R \frac{2x}{\sqrt{R^2 - x^2}}\, dx$$

$$= -\pi R \int_{R^2}^0 u^{-\frac{1}{2}}\, du \qquad \text{(where } u = R^2 - x^2)$$

$$= -2\pi R u^{\frac{1}{2}} \Big|_{R^2}^0$$

$$= 2\pi R^2$$

As this is just the top half of the sphere we must double it to get the correct formula, $4\pi R^2$.

Problem 86 *Find the area of the portion of the plane $z = 2x + 3y$ that lies above the square with vertices at $(0, 0, 0)$, $(1, 0, 0)$, $(0, 1, 0)$, and $(1, 1, 0)$.*

Problem 87 *Find the area of the portion of the graph of $z = \sqrt{x^2 + y^2}$ that lies above the domain*

1. $R = \{(x, y) | 0 \le x \le 1, 0 \le y \le 1\}$
2. $D = \{(x, y) | x^2 + y^2 \le 1\}$

Problem 88 *Derive a formula for the graph of the spherical equation $\rho = f(\phi)$.*

Problem 89 *Find the area of the surface pictured below, parameterized by*

$$\Psi(r, \theta) = (r \cos \theta, r \sin \theta, \theta)$$

$$-1 \le r \le 1, \quad 0 \le \theta \le \pi$$

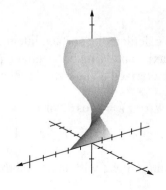

8.5 Surface Integrals

Just as we could integrate a function over a parameterized curve, we can also integrate functions over parameterized surfaces. The resulting integrals are called *surface integrals*. Here's a way to think about surface integrals. Imagine that there is a fog permeating all of $\mathbb{R}^3$. As the fog may not be uniform there is some function $f(x, y, z)$ which gives the density of the water droplets at each point. Now imagine we have some surface in $\mathbb{R}^3$ cutting through the fog. We wish to find out the total amount of water droplets in the surface. If $f(x, y, z) = 1$, then this will just be the area of the surface. If not, then the answer will be the surface integral of f.

Recall the relationship between the arc length of a curve C parameterized by a function $\Psi(t)$ and the line integral of a function $f(x, y)$ over C:

Arc Length	Line Integral
$\int\limits_{a}^{b} \left| \frac{\partial \Psi}{\partial t} \right| dt$	$\int\limits_{a}^{b} f(\Psi(t)) \left| \frac{\partial \Psi}{\partial t} \right| dt$

The relationship between the surface area of a surface S parameterized by a function $\Psi(u, v)$ and the surface integral of a function $f(x, y, z)$ over S is the same:

Surface Area	Surface Integral
$\iint\limits_{R} \left| \frac{\partial \Psi}{\partial u} \times \frac{\partial \Psi}{\partial v} \right| du\, dv$	$\iint\limits_{R} f(\Psi(u, v)) \left| \frac{\partial \Psi}{\partial u} \times \frac{\partial \Psi}{\partial v} \right| du\, dv$

The surface integral of $f(x, y, z)$ over the surface S is denoted as

$$\int\limits_{S} f(x, y, z)\, d\mathbf{S}$$

EXAMPLE 8-10

Let $f(x, y, z) = \frac{z}{x^2+y^2}$. We calculate the surface integral of f over the cylinder of radius 1, centered on the z-axis, and lying between the plane $z = 0$ and $z = 1$.

First, the surface is parameterized using cylindrical coordinates as follows:

$$\Psi(\theta, z) = (\cos\theta, \sin\theta, z)$$

$$0 \le \theta \le 2\pi, \quad 0 \le z \le 1$$

To compute the surface integral we will next need the partial derivatives of this parameterization:

$$\frac{\partial\Psi}{\partial\theta} = \langle -\sin\theta, \cos\theta, 0 \rangle$$

$$\frac{\partial\Psi}{\partial z} = \langle 0, 0, 1 \rangle$$

The cross product of these vectors is

$$\frac{\partial\Psi}{\partial\theta} \times \frac{\partial\Psi}{\partial z} = \begin{vmatrix} \mathbf{i} & \mathbf{j} & \mathbf{k} \\ -\sin\theta & \cos\theta & 0 \\ 0 & 0 & 1 \end{vmatrix} = \langle \cos\theta, \sin\theta, 0 \rangle$$

Finally, the magnitude of the cross product is then

$$\left| \frac{\partial\Psi}{\partial\theta} \times \frac{\partial\Psi}{\partial z} \right| = \sqrt{\cos^2\theta + \sin^2\theta} = 1$$

We are now ready to compute the surface integral:

$$\int_S f(x, y, z)\, d\mathbf{S} = \int_0^1 \int_0^{2\pi} f(\Psi(\theta, z)) \left| \frac{\partial\Psi}{\partial\theta} \times \frac{\partial\Psi}{\partial z} \right| d\theta\, dz$$

$$= \int_0^1 \int_0^{2\pi} \frac{z}{\cos^2\theta + \sin^2\theta} \cdot 1\, d\theta\, dz$$

$$= \int_0^1 \int_0^{2\pi} z\, d\theta\, dz$$

$$= \pi$$

Problem 90 *Calculate the surface integral of the function* $f(x, y, z) = x + y + z$ *over the portion of the plane* $z = x + y$ *that lies above the square in the* xy*-plane where* $0 \leq x \leq 1$ *and* $0 \leq y \leq 1$.

Problem 91 *Calculate the surface integral of the function* $f(x, y, z) = z$ *over the top half of the sphere of radius 1.*

8.6 Volume

In this section, we revisit the calculation of volumes of regions in $\mathbb{R}^3$. If the region is nice enough, we can just compute a triple integral of the function $f(x, y, z) = 1$ over the region. However, if the region is complicated we may want to employ a parameterization to help calculate its volume. The steps to define the volume by way of integrating a parameterization are similar to those used to define arc length and surface area. To this end, we suppose $\Psi(u, v, w)$ is a parameterization of a volume V in $\mathbb{R}^3$, with domain Q. We now derive a formula for the volume of V by executing the following steps:

1. Choose a lattice of points $\{(u_i, v_j, w_k)\}$ in Q. This defines a lattice $\{\Psi(u_i, v_j, w_k)\}$ in V.
2. Connecting adjacent lattice points breaks up V into a bunch of parallelepipeds.
3. We compute the volume of each parallelepiped. The edges are given by the vectors

$$V_u = \Psi(u_{i+1}, v_j, w_k) - \Psi(u_i, v_j, w_k)$$
$$V_v = \Psi(u_i, v_{j+1}, w_k) - \Psi(u_i, v_j, w_k)$$
$$V_w = \Psi(u_i, v_j, w_{k+1}) - \Psi(u_i, v_j, w_k)$$

The desired volume is then given by the determinant of the 3×3 matrix whose rows are V_u, V_v, and V_w. We denote this as

$$|V_u V_v V_w|$$

The usual algebraic trickery transforms this as follows:

$$|V_u V_v V_w| = |V_u V_v V_w| \frac{\Delta u \Delta v \Delta w}{\Delta u \Delta v \Delta w}$$

$$= \left| \frac{V_u}{\Delta u} \frac{V_v}{\Delta v} \frac{V_w}{\Delta w} \right| \Delta u \Delta v \Delta w$$

4. The desired volume is now the limit, as Δu, Δv, and Δw tend toward 0, of the sum of this quantity over all i, j, and k:

$$\text{Volume} = \lim_{\Delta u, \Delta v, \Delta w \to 0} \sum_{i,j,k} \left| \frac{V_u}{\Delta u} \ \frac{V_v}{\Delta v} \ \frac{V_w}{\Delta w} \right| \Delta u \Delta v \Delta w$$

But, as in the computation of surface area,

$$\lim_{\Delta u \to 0} \frac{V_u}{\Delta u} = \lim_{\Delta u \to 0} \frac{\Psi(u_{i+1}, v_j, w_k) - \Psi(u_i, v_j, w_k)}{\Delta u} = \frac{\partial \Psi}{\partial u}$$

Hence, after taking the limit the above summation becomes

$$\text{Volume} = \int \int_Q \int \left| \frac{\partial \Psi}{\partial u} \ \frac{\partial \Psi}{\partial v} \ \frac{\partial \Psi}{\partial w} \right| du \, dv \, dw$$

EXAMPLE 8-11

We show that the volume bounded by a sphere of radius R is $\frac{4}{3}\pi R^3$. One way to do this is to note that the top half of the sphere can be realized as the graph of the equation $z = \sqrt{R^2 - x^2 - y^2}$. The desired volume can thus be obtained by doubling a double integral:

$$\text{Volume} = 2 \int_{-1}^{1} \int_{-\sqrt{R^2-y^2}}^{\sqrt{R^2-x^2}} \sqrt{R^2 - x^2 - y^2} \, dx \, dy$$

However, this integral is difficult to evaluate.

Instead, we employ a parameterization. The sphere is best parameterized in the usual way by using spherical coordinates:

$$\Psi(\rho, \theta, \phi) = (\rho \sin \phi \cos \theta, \rho \sin \phi \sin \theta, \rho \cos \phi)$$

$$0 \le \rho \le R, \quad 0 \le \theta \le 2\pi, \quad 0 \le \phi \le \pi$$

We now compute all three partials of the parameterization:

$$\frac{\partial \Psi}{\partial \rho} = \langle \sin \phi \cos \theta, \sin \phi \sin \theta, \cos \phi \rangle$$

$$\frac{\partial \Psi}{\partial \theta} = \langle -\rho \sin \phi \sin \theta, \rho \sin \phi \cos \theta, 0 \rangle$$

$$\frac{\partial \Psi}{\partial \phi} = \langle \rho \cos \phi \cos \theta, \rho \cos \phi \sin \theta, -\rho \sin \phi \rangle$$

The integrand that we will need is the determinant of the 3×3 matrix whose rows are the above vectors:

$$\left| \frac{\partial \Psi}{\partial \rho} \frac{\partial \Psi}{\partial \theta} \frac{\partial \Psi}{\partial \phi} \right| = \begin{vmatrix} \sin\phi\cos\theta & \sin\phi\sin\theta & \cos\phi \\ -\rho\sin\phi\sin\theta & \rho\sin\phi\cos\theta & 0 \\ \rho\cos\phi\cos\theta & \rho\cos\phi\sin\theta & -\rho\sin\phi \end{vmatrix}$$

$$= -\rho^2 \sin\phi \quad \text{(Check this!)}$$

We now get the desired volume by integrating:

$$\text{Volume} = \int_0^\pi \int_0^{2\pi} \int_0^R \left| \frac{\partial \Psi}{\partial \rho} \frac{\partial \Psi}{\partial \theta} \frac{\partial \Psi}{\partial \phi} \right| d\rho \, d\theta \, d\phi$$

$$= \int_0^\pi \int_0^{2\pi} \int_0^R -\rho^2 \sin\phi \, d\rho \, d\theta \, d\phi$$

$$= \int_0^\pi \int_0^{2\pi} \left. \frac{-\rho^3 \sin\phi}{3} \right|_{\rho=0}^R d\theta \, d\phi$$

$$= \int_0^\pi \int_0^{2\pi} \frac{-R^3 \sin\phi}{3} \, d\theta \, d\phi$$

$$= \frac{2\pi R^3}{3} \int_0^\pi -\sin\phi \, d\phi$$

$$= \frac{2\pi R^3}{3} \cos\phi|_0^\pi$$

$$= \frac{-4\pi R^3}{3}$$

The negative sign is an unfortunate artifact of the fact that the determinant of a matrix will change sign if you swap two rows or two columns. We will see in later sections that this can be used to our advantage. For now, though, just note that a volume must always be positive. Hence, we take the absolute value of the above answer to recover the correct formula, $\frac{4}{3}\pi R^3$.

Although this method may seem very complicated at first, observe that it has a significant virtue: each step in the integration was very easy. This is the advantage of using parameterizations to compute volumes. Any volume problem can be solved by a double or triple integral. But employing a parameterization may transform difficult integrals into easy ones.

Problem 92 *Show that the volume bounded by a cylinder of radius R and height h is $\pi R^2 h$ by integrating with a parameterization from cylindrical coordinates.*

Problem 93 *Calculate the volume under the cone $z = 1 - x^2 - y^2$, and above the xy-plane by employing the parameterization*

$$\Psi(r, \theta, z) = ((1 - z)r \cos \theta, (1 - z)r \sin \theta, z)$$

$$0 \leq r \leq 1, \quad 0 \leq \theta \leq 2\pi, \quad 0 \leq z \leq 1$$

8.7 Change of Variables

In Calculus I and II you learn how to integrate a function of one variable over a domain in $\mathbb{R}^1$. In Chapter 4, we learned how to integrate a function of two variables over a domain in $\mathbb{R}^2$, and a function of three variables over a domain in $\mathbb{R}^3$:

$$\int \int_V \int f(x, y, z) \, dx \, dy \, dz$$

If the domain V has a convenient parameterization then it may be easier to use this than to directly perform the integral. This is very similar to the relationship between line integrals and arc length, or surface integrals and surface area. As usual, we let Ψ denote a parameterization of V, with domain Q. If V is two dimensional we have

$$\int \int_V f(x, y) \, dx \, dy = \int \int_Q f(\Psi(u, v)) \left| \frac{\partial \Psi}{\partial u} \frac{\partial \Psi}{\partial v} \right| du \, dv$$

And if V is three dimensional

$$\int \int_V \int f(x, y, z) \, dx \, dy \, dz = \int \int_Q \int f(\Psi(u, v, w)) \left| \frac{\partial \Psi}{\partial u} \frac{\partial \Psi}{\partial v} \frac{\partial \Psi}{\partial w} \right| du \, dv \, dw$$

EXAMPLE 8-12

Recall the functions

$$\cosh t = \frac{e^t + e^{-t}}{2}, \quad \sinh t = \frac{e^t - e^{-t}}{2}$$

These functions conveniently satisfy

$$\cosh^2 t - \sinh^2 t = 1$$

$$\frac{d}{dt} \cosh t = \sinh t, \quad \frac{d}{dt} \cosh t = \sinh t$$

The first of these identities means that $(\cosh t, \sinh t)$ are the coordinates of a point on the hyperbola $x^2 - y^2 = 1$. Suppose, we want to find the area of the region pictured below.

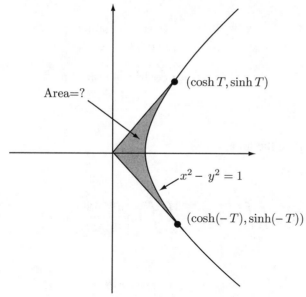

This region is parameterized by

$$\Psi(r, t) = (r \cosh t, r \sinh t)$$

$$0 \le r \le 1, \quad -T \le t \le T$$

The area of a region R of $\mathbb{R}^2$ is computed by integrating the function $f(x, y) = 1$:

$$\text{Area} = \iint_R 1 \, dx \, dy$$

However, to calculate this integral directly would entail finding complicated limits of integration. Instead we use the parameterization Ψ. To do this, we will need the partial derivatives of Ψ:

$$\frac{\partial \Psi}{\partial r} = \langle \cosh t, \sinh t \rangle$$

$$\frac{\partial \Psi}{\partial t} = \langle r \sinh t, r \cosh t \rangle$$

The determinant of the matrix of partials is thus

$$\left| \frac{\partial \Psi}{\partial r} \; \frac{\partial \Psi}{\partial t} \right| = \left| \begin{matrix} \cosh t & \sinh t \\ r \sinh t & r \cosh t \end{matrix} \right|$$

$$= r \cosh^2 t - r \sinh^2 t$$

$$= r$$

We now compute the area:

$$\iint\limits_{R} 1 \, dx \, dy = \int\limits_{-T}^{T} \int\limits_{0}^{1} 1 \left| \frac{\partial \Psi}{\partial r} \; \frac{\partial \Psi}{\partial t} \right| dr \, dt$$

$$= \int\limits_{-T}^{T} \int\limits_{0}^{1} r \, dr \, dt$$

$$= \int\limits_{-T}^{T} \left. \frac{r^2}{2} \right|_{0}^{1} dt$$

$$= \frac{1}{2} \int\limits_{-T}^{T} dt$$

$$= T$$

EXAMPLE 8-13

We evaluate the integral

$$\int_0^1 \int_0^1 \int_0^{\sqrt{1-y^2}} x^2 + y^2 + z \, dx \, dy \, dz$$

Doing this directly is difficult, so we employ a parameterization. The domain of integration is the intersection of a cylinder of height 1 and radius 1, centered on the z-axis, with the positive octant. This can be parameterized with cylindrical coordinates by

$$\Psi(r, \theta, z) = (r\cos\theta, r\sin\theta, z)$$

$$0 \le r \le 1, \quad 0 \le \theta \le \tfrac{\pi}{2}, \quad 0 \le z \le 1$$

The derivatives of this parameterization are

$$\frac{\partial\Psi}{\partial r} = \langle \cos\theta, \sin\theta, 0 \rangle$$

$$\frac{\partial\Psi}{\partial\theta} = \langle -r\sin\theta, r\cos\theta, 0 \rangle$$

$$\frac{\partial\Psi}{\partial z} = \langle 0, 0, 1 \rangle$$

To do the integral we will need the determinant of the 3×3 matrix whose rows are these vectors:

$$\begin{vmatrix} \cos\theta & \sin\theta & 0 \\ -r\sin\theta & r\cos\theta & 0 \\ 0 & 0 & 1 \end{vmatrix} = r$$

We denote the function $x^2 + y^2 + z$ in the original integral by $f(x, y, z)$. Then

$$\int_0^1 \int_0^1 \int_0^{\sqrt{1-y^2}} x^2 + y^2 + z \, dx \, dy \, dz = \int_0^1 \int_0^{\frac{\pi}{2}} \int_0^1 f(\Psi(r, \theta, z)) \left| \frac{\partial\Psi}{\partial r} \frac{\partial\Psi}{\partial\theta} \frac{\partial\Psi}{\partial z} \right| dr \, d\theta \, dz$$

$$= \int_0^1 \int_0^{\frac{\pi}{2}} \int_0^1 ((r\cos\theta)^2 + (r\sin\theta)^2 + z)r \, dr \, d\theta z$$

$$= \int_0^1 \int_0^{\frac{\pi}{2}} \int_0^1 r^3 + rz \, dr \, d\theta \, dz$$

$$= \int_0^1 \int_0^{\frac{\pi}{2}} \frac{1}{4} r^4 + \frac{1}{2} r^2 z \Big|_0^1 d\theta \, dz$$

$$= \int_0^1 \int_0^{\frac{\pi}{2}} \frac{1}{4} + \frac{1}{2} z \, d\theta \, dz$$

$$= \int_0^1 \frac{\pi}{8} + \frac{\pi}{4} z \, dz$$

$$= \frac{\pi}{8} z + \frac{\pi}{8} z^2 \Big|_0^1$$

$$= \frac{\pi}{4}$$

Problem 94 *Calculate the integral of the function $f(x, y) = 2x - y$ over the elliptical region in $\mathbb{R}^2$ parameterized by*

$$\Psi(r, \theta) = (2r \cos \theta, r \sin \theta)$$

$$0 \le r \le 1, \quad 0 \le \theta \le \pi$$

Problem 95 *Calculate the integral of the function $f(x, y, z) = \frac{1}{1+z^2}$ over the region V of $\mathbb{R}^3$ parameterized by*

$$\Psi(r, \theta, \omega) = (r \cosh \omega \cos \theta, r \cosh \omega \sin \theta, \sinh \omega)$$

$$1 \le r \le 2, \quad 0 \le \theta \le 2\pi, \quad -2 \le \omega \le 2$$

(The region V is between two hyperboloids, as depicted below.)

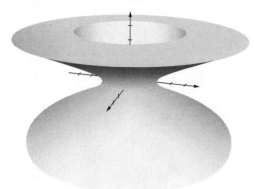

Problem 96 *Let V be the region inside the cylinder of radius 1, centered around the z-axis, and between the planes $z = 0$ and $z = 2$. Integrate the function $f(x, y, z) = z$ over V.*

Problem 97 *Let R be the region in the first quadrant of $\mathbb{R}^2$, below the line $y = x$, and bounded by $x^2 + y^2 = 4$. Integrate the function*

$$f(x, y) = 1 + \frac{y^2}{x^2}$$

over R.

Quiz

Problem 98

1. a. *Find a parameterization for the set of points in $\mathbb{R}^2$ that satisfies the equation $x = \sin y$.*

   b. *Find a unit tangent vector to this curve at the point $(0, 0)$.*

2. *Let Q be the region under the graph of $f(x, y) = x^2 + y^2$ and above the square with vertices at $(0, 0, 0)$, $(1, 0, 0)$, $(0, 1, 0)$, and $(1, 1, 0)$.*

   a. *Find a parameterization for Q of the form $\Psi(x, y, t)$, where $0 \leq x, y, t \leq 1$.*

   b. *Use the parameterization Ψ to integrate over Q the function*

   $$f(x, y, z) = \frac{z}{x^2 + y^2}$$

3. *Let R be the region of $\mathbb{R}^2$ parameterized by*

   $$\phi(r, t) = (r \cosh t, r \sinh t), \quad 0 \leq r \leq 1, -1 \leq t \leq 1$$

   Integrate the function $f(x, y) = x^2 - y^2$ over R.

CHAPTER 9

Vector Fields and Derivatives

9.1 Definition

A *vector field* is simply a choice of vector for each point. So, for example, a vector field on $\mathbb{R}^2$ would have some vector at the point $(1, 2)$, some other vector at the point $(-1, 1)$, etc. We often draw vector fields by picking a few points and drawing the vector based at those points.

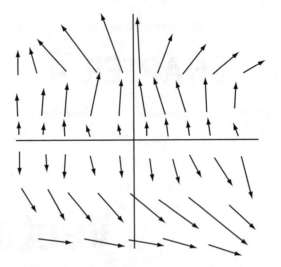

More formally, a vector field is a function from $\mathbb{R}^2$ to $\mathbb{R}^2$. What goes in to this function are the coordinates of the point where you are. What comes out are the components of the vector at that point.

EXAMPLE 9-1

Consider the vector field $\langle xy, x + y \rangle$. At the point $(1, 1)$ this vector field contains the vector $\langle 1, 2 \rangle$. At the point $(2, 4)$ it contains the vector $\langle 8, 6 \rangle$. If we use a computer to draw it we would see something like

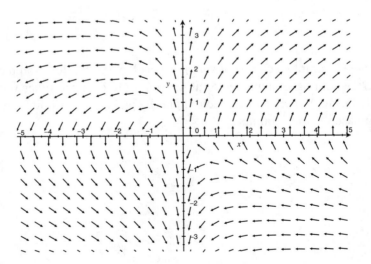

EXAMPLE 9-2

The vector field $\langle \sin x, \sin y \rangle$ is depicted below.

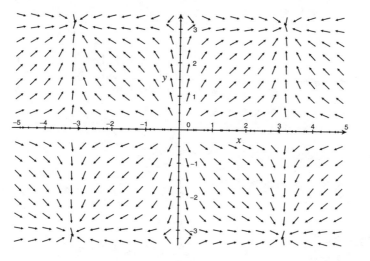

Problem 99 *Sketch the following vector fields.*

1. $\langle x, y \rangle$
2. $\langle x, -y \rangle$
3. $\langle y, -x \rangle$

9.2 Gradients, Revisited

We have already seen many vector fields, although we did not use this language. Whenever we take a function f and compute its gradient ∇f at a point we get a vector. The set of all such vectors is then a vector field, which we now call "grad f."

EXAMPLE 9-3

Suppose $f(x, y) = xy^2$. Then the gradient of $f(x, y)$ at the point (x, y) is $\nabla f(x, y) = \langle y^2, 2xy \rangle$. If we draw this vector at various values of x and y, we get the picture depicted below.

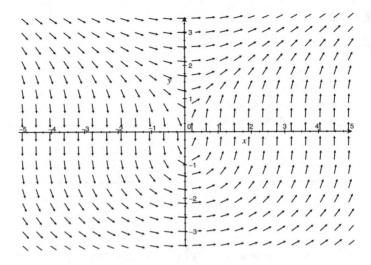

EXAMPLE 9-4

Let $f(x, y, z) = xy^2z^3$. Then $\nabla f = \langle y^2z^3, 2xyz^3, 3xy^2z^2 \rangle$.

Problem 100 *For each of the following functions, f, compute ∇f.*

1. $f(x, y) = x + y$
2. $f(x, y, z) = x + yz$
3. $f(x, y, z) = xy + xz + yz$

9.3 Divergence

In the previous section, we saw that the gradient operator gives us a way to take a function $f(x, y)$ and get a vector field. In this section, we explore a way to take a vector field and get a function. Eventually, we will see that the value of this function at a point is a measure of how much the vector field is "spreading out" there.

Suppose $\langle f(x, y, z), g(x, y, z), h(x, y, z) \rangle$ is a vector field $\mathbf{V}$ on $\mathbb{R}^3$. Then we define the *divergence* of $\mathbf{V}$, "Div $\mathbf{V}$," to be the function

$$\frac{\partial f}{\partial x} + \frac{\partial g}{\partial y} + \frac{\partial h}{\partial z}$$

Note that the first term is associated with the first component of $\mathbf{V}$, the second term with the second component, and the third term with the third component. This, and the fact that the terms are being added, should remind you of the dot product. This gives us a purely notational way to remember how to calculate the divergence of a vector field. We let ∇ denote the "vector" $\left\langle \frac{\partial}{\partial x}, \frac{\partial}{\partial y}, \frac{\partial}{\partial z} \right\rangle$. This, of course, is only a vector in a notational sense. But if we suspend our disbelief for a moment and allow such absurdities, we can write the formula for the divergence of a vector field in a very compact way:

$$\text{Div } \mathbf{V} = \nabla \cdot \mathbf{V}$$

EXAMPLE 9-5
Let $\mathbf{V}$ be the vector field $\langle x^2 y, x + yz, xy^2 z^3 \rangle$. Then the divergence of $\mathbf{V}$ is calculated as follows:

$$\text{Div } \mathbf{V} = \nabla \cdot \mathbf{V}$$

$$= \left\langle \frac{\partial}{\partial x}, \frac{\partial}{\partial y}, \frac{\partial}{\partial z} \right\rangle \cdot \langle x^2 y, x + yz, xy^2 z^3 \rangle$$

$$= \frac{\partial}{\partial x}(x^2 y) + \frac{\partial}{\partial y}(x + yz) + \frac{\partial}{\partial z}(xy^2 z^3)$$

$$= 2xy + z + 3xy^2 z^2$$

Problem 101 *Compute the divergence of the following vector fields.*

1. $\langle y, z, x \rangle$
2. $\langle x + y, x - y, z \rangle$
3. $\langle x^2 + y^2, x^2 - y^2, z^2 \rangle$

Problem 102 *Let $f(x, y, z)$ be a function. What is $\nabla \cdot \nabla f$?*

Problem 103 *Show that*

$$\nabla \cdot (\mathbf{F} \times \mathbf{G}) = \mathbf{G} \cdot (\nabla \times \mathbf{F}) - \mathbf{F} \cdot (\nabla \times \mathbf{G})$$

9.4 Curl

There are many useful ways to apply partial derivatives to vector fields. We have already seen that gradients give us a way to take a function and get a vector field. Then we saw that divergence is a way to take a vector field and get a function. Here, we define a way to use partial derivatives to transform one vector field into another.

First, recall our notational absurdity, $\nabla = \left\langle \frac{\partial}{\partial x}, \frac{\partial}{\partial y}, \frac{\partial}{\partial z} \right\rangle$. In the previous section, we defined a new operation by using ∇ in a dot product. Here, we define an operation called *curl* by using ∇ in a cross product. As before, let $\mathbf{V} = \langle f(x, y, z), g(x, y, z), h(x, y, z) \rangle$. Then we define

$$\text{curl } \mathbf{V} = \nabla \times \mathbf{V}$$

$$= \begin{vmatrix} \mathbf{i} & \mathbf{j} & \mathbf{k} \\ \frac{\partial}{\partial x} & \frac{\partial}{\partial y} & \frac{\partial}{\partial z} \\ f & g & h \end{vmatrix}$$

$$= \left(\frac{\partial h}{\partial y} - \frac{\partial g}{\partial z} \right) \mathbf{i} - \left(\frac{\partial h}{\partial x} - \frac{\partial f}{\partial z} \right) \mathbf{j} + \left(\frac{\partial g}{\partial x} - \frac{\partial f}{\partial y} \right) \mathbf{k}$$

$$= \left\langle \frac{\partial h}{\partial y} - \frac{\partial g}{\partial z}, \frac{\partial f}{\partial z} - \frac{\partial h}{\partial x}, \frac{\partial g}{\partial x} - \frac{\partial f}{\partial y} \right\rangle$$

EXAMPLE 9-6
Let $\mathbf{V}$ be the vector field $\langle x^2 y, x + yz, xy^2 z^3 \rangle$. Then the curl of $\mathbf{V}$ is calculated as follows:

$$\text{curl } \mathbf{V} = \nabla \times \mathbf{V}$$

$$= \begin{vmatrix} \mathbf{i} & \mathbf{j} & \mathbf{k} \\ \frac{\partial}{\partial x} & \frac{\partial}{\partial y} & \frac{\partial}{\partial z} \\ x^2 y & x + yz & xy^2 z^3 \end{vmatrix}$$

$$= \langle 2xyz^3 - y, -y^2 z^3, 1 - x^2 \rangle$$

Later we will see that the curl of a vector field measures how much it "twists" at each point.

Problem 104 *Find the curl of each of the following vector fields.*

    1. $\langle x + y, x - z, y + z \rangle$
    2. $\langle yz, -xz, xy \rangle$

Problem 105 *Let $f(x, y, z)$ be a function and $\mathbf{V}$ a vector field on $\mathbb{R}^3$. Compute*

    1. $\nabla \times (\nabla f)$
    2. $\nabla \cdot (\nabla \times \mathbf{V})$

Quiz

Problem 106

1. *Find the curl of the following vector field:*

$$\left\langle \frac{-y}{x^2 + y^2}, \frac{x}{x^2 + y^2}, 0 \right\rangle$$

2. *Find the divergence of the following vector field:*

$$\left\langle x^2 + y^2, y^2 - x^2, 0 \right\rangle$$

3. *Find the gradient of*

$$f(x, y, z) = x^2 \sin(y - z)$$

4. *Find two vector fields whose curls are* $\left\langle 0, 0, \frac{y}{x} \right\rangle$.

CHAPTER 10

Integrating Vector Fields

In the previous chapter we introduced the concept of a vector field, and looked at various concepts of differentiation in relation to these objects. In this chapter, we look at ways to integrate a vector field. From calculus you should suspect that when we combine integration and differentiation something amazing will happen, akin to the *Fundamental Theorem of Calculus*. This is the topic of the next chapter.

10.1 Line Integrals

We will always integrate a vector field over a domain that can be parameterized by a line, surface, or volume. In this section, we look at the first of these situations, integrals over parameterized curves. As in Section 8.3, these will be called *line integrals*. However, the definition of the integral of a vector field over a parameterized curve is a bit different than the definition of a function over the same curve, so there is potential for confusion here.

Before giving the definition, it may help to give a physical motivation for line integrals. Drop a leaf into a flowing stream and follow it. Eventually, you may see it encounter a stick caught on some rocks. If the stick is perpendicular to the flow of water, the leaf will get stuck. If the stick is parallel to the flow, then the leaf rushes by as the force of the flowing water pushes it along. But if the stick is at some angle to the flow the leaf may hit it, and then slowly work its way past.

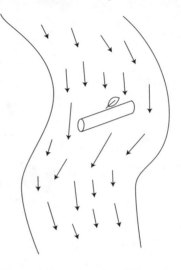

One can explain this by looking at the force the water can exert on the leaf. The leaf can only travel in a direction parallel to the stick, which we can represent as a vector, L. The water flowing under the stick exerts some force on the leaf, which can also be represented as a vector, W. If these vectors are perpendicular then the leaf does not move, so the net force F on the leaf must be zero. On the other hand, the leaf moves the fastest when L and W are parallel, so F must be greatest in this situation. It should be no surprise, then, that F is proportional to the dot product, $L \cdot W$.

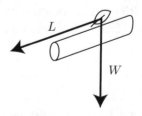

Before proceeding further, we must dispense with an important technicality. In the above figure, there were two choices for the vector L. One of these is depicted in the figure, and the other points in the exact opposite direction (but still along the stick). If the other choice was made, then the value of $L \cdot W$ would have exactly the opposite sign. Which is the correct choice? There is no right answer to this. We just have to declare, before we do any problem, which is correct. Such a declaration is called an *orientation*. We will say more about this idea shortly. For now, just note that if the wrong choice was made then we can correct things by changing the sign of our final answer.

Now we suppose that somehow the leaf is confined to some parameterized curve, C, as the water rushes past. Sometimes the water may be perpendicular to the curve, and sometimes it may be parallel. The problem is to evaluate the total force that the water exerts on the leaf as it travels along the curve. (If the leaf always moves with unit speed then this is called the *work* done by the water.)

Let $\Psi(t)$ be a parameterization of C. Let $\mathbf{W}$ be a vector field that gives the direction and speed of the water at each point of the stream. The direction that the leaf is moving at the point $\Psi(t)$ is the vector $\frac{d\Psi}{dt}$. Let $\mathbf{W}(\Psi(t))$ be the vector of the vector field $\mathbf{W}$ that is based at the point $\Psi(t)$. Then the quantity we want to evaluate is

$$\int \mathbf{W}(\Psi(t)) \cdot \frac{d\Psi}{dt} \, dt$$

This integral is called the *line integral* of $\mathbf{W}$ over C, and is often denoted as

$$\int_C \mathbf{W} \cdot d\mathbf{s}$$

Our answer, unfortunately, is still incomplete. Recall that at some point we must make a choice, called an orientation. This amounts to deciding if the vector $\frac{d\Psi}{dt}$ points the "right way" or the "wrong way." An orientation can be denoted pictorially just by an arrow along C, showing the correct choice. But this does not mean that to do the integral, we must choose a parameterization whose derivative points the right way. If it does not then we just need to change the sign of our final answer.

EXAMPLE 10-1
Let $\mathbf{W} = \langle xy, x + y \rangle$. Let C be the oriented curve depicted below.

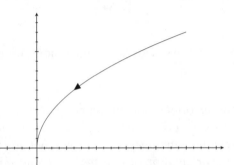

The orientation is given by the direction of the arrow in the figure. The curve is parameterized by

$$\Psi(t) = (t^2, 3t), \quad 0 \le t \le 2$$

At the point $\Psi(1)$ one can easily check that $\frac{d\Psi}{dt} = \langle 2, 3 \rangle$. This vector points in the direction opposite to the orientation. This just means that after we integrate we will have to flip the sign of our final answer.

We now calculate the line integral of $\mathbf{W}$ along C using Ψ. First, note that

$$\mathbf{W}(\Psi(t)) = \langle (t^2)(3t), t^2 + 3t \rangle = \langle 3t^3, t^2 + 3t \rangle$$

and

$$\frac{d\Psi}{dt} = \langle 2t, 3 \rangle$$

$$\int \mathbf{W}(\Psi(t)) \cdot \frac{d\Psi}{dt} \, dt = \int_0^2 \langle 3t^3, t^2 + 3t \rangle \cdot \langle 2t, 3 \rangle \, dt$$

$$= \int_0^2 6t^4 + 3t^2 + 9t \, dt$$

$$= \frac{6}{5}t^5 + t^3 + \frac{9}{2}t^2 \Big|_0^2$$

$$= 64\frac{2}{5}$$

As the orientation disagrees with Ψ we must now flip the sign, obtaining our final answer, $-64\frac{2}{5}$.

An easy way to give an orientation on a curve parameterized by a function $\Psi(t)$ is to simply declare that $\frac{d\Psi}{dt}$ points in the right direction. In this case, we may go ahead and use Ψ to evaluate the integral of $\mathbf{W}$ without any further considerations. In this case, we say the orientation of C is the one *induced* by Ψ.

EXAMPLE 10-2
Let C be the curve parameterized by

$$\Psi(t) = (\cos t, t^2), \quad 0 \le t, \le \frac{\pi}{4}$$

with the orientation induced by Ψ.

Let $\mathbf{W} = \langle \frac{1}{x}, 0 \rangle$. We integrate $\mathbf{W}$ over C.

First, note that

$$\mathbf{W}(\Psi(t)) = \left\langle \frac{1}{\cos t}, 0 \right\rangle$$

Next, observe

$$\frac{d\Psi}{dt} = \langle -\sin t, 2t \rangle$$

We now integrate

$$\int_C \mathbf{W} \cdot d\mathbf{s} = \int_0^{\frac{\pi}{4}} \left\langle \frac{1}{\cos t}, 0 \right\rangle \cdot \langle -\sin t, 2t \rangle \ dt$$

$$= \int_0^{\frac{\pi}{4}} -\tan t \ dt$$

$$= -\ln|\sec t| \Big|_0^{\frac{\pi}{4}}$$

$$= -\ln\sqrt{2}$$

As the orientation of C is chosen to agree with Ψ we do not need to worry about changing the sign of our answer.

One may wonder about the connection between line integrals of vector fields as defined here and line integrals of functions defined previously. Let f denote a function on $\mathbb{R}^n$ and C a curve in $\mathbb{R}^n$ parameterized by $\Psi(t)$. Let $\mathbf{W}$ be a vector field chosen so that at the point $\Psi(t)$ of C the vector $\mathbf{W}(\Psi(t))$ is tangent to C (i.e., it points in the same direction as $\frac{d\Psi}{dt}$) and has length $f(\Psi(t))$. Then we have

$$\mathbf{W}(\Psi(t)) = f(\Psi(t)) \frac{\frac{d\Psi}{dt}}{\left|\frac{d\Psi}{dt}\right|}$$

The definition of the line integral of $\mathbf{W}$ over C then gives

$$\int_C \mathbf{W} \cdot d\mathbf{s} = \int \mathbf{W}(\Psi(t)) \cdot \frac{d\Psi}{dt}\, dt$$

$$= \int f(\Psi(t)) \frac{\frac{d\Psi}{dt}}{\left|\frac{d\Psi}{dt}\right|} \cdot \frac{d\Psi}{dt}\, dt$$

$$= \int \frac{f(\Psi(t))}{\left|\frac{d\Psi}{dt}\right|} \frac{d\Psi}{dt} \cdot \frac{d\Psi}{dt}\, dt$$

$$= \int \frac{f(\Psi(t))}{\left|\frac{d\Psi}{dt}\right|} \left|\frac{d\Psi}{dt}\right|^2 dt$$

$$= \int f(\Psi(t)) \left|\frac{d\Psi}{dt}\right| dt$$

This last equality is precisely the definition of the line integral of f over C.

Problem 107 *Calculate the line integral of $\langle -y, x \rangle$ over the circle of radius 1, oriented counterclockwise.*

Problem 108 *Let $\mathbf{W} = \langle xy, xz^2, y + z \rangle$. Let C be the curve parameterized by*

$$\Psi(t) = (t^2, t, 1 - t), \quad 0 \le t \le 1$$

with the induced orientation. Calculate the line integral of $\mathbf{W}$ over C.

Problem 109 *Let $f(x, y) = xy^2$. Let C be the portion of the parabola $y = x^2$ parameterized by*

$$\Psi(t) = (t, t^2), \quad -1 \le t \le 2$$

with the induced orientation.

1. Calculate the integral of the vector field ∇f over C.
2. Calculate $f(\Psi(2)) - f(\Psi(-1))$.

Problem 110 *Let C be the subset of the graph of $y = x^2$ where $0 \leq x \leq 1$ (oriented away from the origin). Let $\mathbf{W}$ be the vector field $\langle -x^4, xy \rangle$. Integrate $\mathbf{W}$ over C.*

10.2 Surface Integrals

In Section 8.3 we defined line integrals of functions, and in the previous section of this chapter we defined line integrals of vector fields. By analogy we will now define surface integrals of vector fields, just as we had defined surface integrals of functions.

The physical motivation for surface integrals of vector fields can again be seen by looking at a river of flowing water. Imagine a net placed vertically in the river. Suppose you wanted to calculate the amount of water flowing through the net at some particular moment in time. Let W denote a vector representing the flow of water at some point of the net. Let N be a vector which is perpendicular to the net. If the net is parallel to the direction of flow, then no water passes through it. In this situation, N is perpendicular to W. The most water passes through the net when the net is perpendicular to the flow. In this case, N and W are parallel. It stands to reason, then, that the amount of water flowing through the net is proportional to $W \cdot N$.

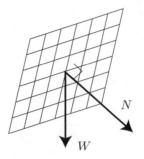

Once again we encounter here a technical difficulty. Why did we draw the vector N in the figure the way we did? If the only condition on N is that it is perpendicular to the net, then why couldn't N point in the exact opposite direction? The answer is that it can! Which choice is "correct" will again depend on a choice, called an *orientation*. We will discuss this more shortly.

Now let S be a surface (representing the net), parameterized by $\Psi(u, v)$. The vectors $\frac{d\Psi}{du}$ and $\frac{d\Psi}{dv}$ are both tangent to S. Hence, the vector $\frac{d\Psi}{du} \times \frac{d\Psi}{du}$ is perpendicular, or *normal*, to S. Let $\mathbf{W}$ be a vector field representing the water flow, and $\mathbf{W}(\Psi(u, v))$ the vector of $\mathbf{W}$ at the point $\Psi(u, v)$. Then the total amount of water flowing through S is given by the integral

$$\int \int \mathbf{W}(\Psi(u, v)) \cdot \left(\frac{d\Psi}{du} \times \frac{d\Psi}{du} \right) \, du \, dv$$

We call this the *surface integral* of $\mathbf{W}$ over S, and denote it by

$$\int_S \mathbf{W} \cdot d\mathbf{S}$$

But this formula isn't the whole story. We still must deal with the issue of orientations. Otherwise, two people evaluating $\int_S \mathbf{W} \cdot d\mathbf{S}$ may get different answers, depending on which parameterization of S they use to evaluate the integral. One way to give an orientation is to say which way is "up" at each point of S. This can be done by giving a vector O which is perpendicular to S at some point. Our parameterization Ψ "agrees" with the choice of orientation if the vector $\frac{d\Psi}{du} \times \frac{d\Psi}{du}$ points in the same direction as O. If our parameterization does not agree with the specified orientation then we can remedy things simply by negating our final answer.

EXAMPLE 10-3
Let $\mathbf{W} = \langle y, x, z \rangle$. Let S be the portion of a cylinder parameterized by

$$\Psi(\theta, z) = (\cos\theta, \sin\theta, z)$$

$$0 \leq \theta \leq \frac{\pi}{4}, \quad 0 \leq z \leq 1$$

The vector $\langle -1, 0, 0 \rangle$ defines an orientation on S at the point $(1, 0, 0)$. We now compute the integral of $\mathbf{W}$ over the surface S with this orientation.

First, we compute the partials,

$$\frac{\partial \Psi}{\partial \theta} = \langle -\sin\theta, \cos\theta, 0 \rangle$$

$$\frac{\partial \Psi}{\partial z} = \langle 0, 0, 1 \rangle$$

And so,

$$\frac{\partial \Psi}{\partial \theta} \times \frac{\partial \Psi}{\partial z} = \begin{vmatrix} \mathbf{i} & \mathbf{j} & \mathbf{k} \\ -\sin\theta & \cos\theta & 0 \\ 0 & 0 & 1 \end{vmatrix} = \langle \cos\theta, \sin\theta, 0 \rangle$$

Now notice that the point $(1, 0, 0) = \Psi(0, 0)$. The vector $\frac{\partial \Psi}{\partial \theta} \times \frac{\partial \Psi}{\partial z}$ is equal to $\langle 1, 0, 0 \rangle$ at this point. This is exactly opposite to the specified orientation of S. Hence, if we use the parameterzation Ψ to compute the integral of $\mathbf{W}$ over S we will have to remember to negate our final answer.

We now integrate

$$\int_S \mathbf{W} \cdot d\mathbf{S} = \int_0^1 \int_0^{\frac{\pi}{4}} \mathbf{W}(\Psi(\theta, z)) \cdot \left(\frac{\partial \Psi}{\partial \theta} \times \frac{\partial \Psi}{\partial z} \right) d\theta \, dz$$

$$= \int_0^1 \int_0^{\frac{\pi}{4}} \langle \sin\theta, \cos\theta, z \rangle \cdot \langle \cos\theta, \sin\theta, 0 \rangle \, d\theta \, dz$$

$$= \int_0^1 \int_0^{\frac{\pi}{4}} 2\sin\theta \cos\theta \, d\theta \, dz$$

$$= \int_0^1 \int_0^{\frac{\pi}{4}} \sin 2\theta \, d\theta \, dz$$

$$= \int_0^1 -\frac{1}{2} \cos 2\theta \Big|_0^{\frac{\pi}{4}} \, dz$$

$$= \int_0^1 \frac{1}{2} \, dz$$

$$= \frac{1}{2}$$

As the parameterization Ψ disagreed with the specified orientation the correct answer is $-\frac{1}{2}$.

Problem 111 *Let* $\mathbf{W} = \langle x, y, z \rangle$. *Integrate* $\mathbf{W}$ *over the unit sphere, with orientation given by the vector* $\langle 1, 0, 0 \rangle$ *at the point* $(1, 0, 0)$.

Problem 112 *Let* $\mathbf{W} = \langle xz, yz, 0 \rangle$. *Integrate* $\mathbf{W}$ *over the surface S paramaterized by*

$$\Psi(u, v) = (u, v, u^2 + v^2)$$

$$0 \leq u \leq 1, \quad 0 \leq v \leq 1$$

Use the induced orientation.

Problem 113 *Let S be the surface given by the following parameterization:*

$$\Psi(\theta, \phi) = (\cos \phi \cos \theta, \cos \phi \sin \theta, \sin \phi)$$

$$0 \leq \theta \leq 2\pi, \quad -\frac{\pi}{4} \leq \phi \leq \frac{\pi}{4}$$

(Note that this is not quite spherical coordinates.)
Integrate over S the vector field

$$\mathbf{W} = \left\langle \frac{1}{x}, \frac{1}{y}, 0 \right\rangle$$

Problem 114 *Let* $\mathbf{W} = \langle xz, yz, 0 \rangle$. *Let S denote the intersection of the unit sphere with the positive octant. An orientation on S is given by the vector* $\langle 0, -1, 0 \rangle$ *at the*

point $(0, 1, 0)$. *Compute*

$$\int_S (\nabla \times \mathbf{W}) \cdot d\mathbf{S}$$

Quiz

Problem 115

1. *Let C be the portion of the graph of $x = y^2$ where $0 \leq x \leq 1$, as depicted in the figure. Integrate the vector field $\langle 1, 1 \rangle$ over C with the indicated orientation.*

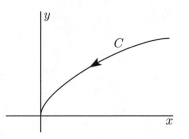

2. *Let S be the frustum parameterized by*

$$\Psi(r, \theta) = (r \cos \theta, r \sin \theta, r),\ 1 \leq r \leq 2, 0 \leq \theta \leq 2\pi$$

Integrate the vector field $\left\langle \frac{1}{x}, -\frac{1}{y}, z \right\rangle$ over S with the induced orientation.

CHAPTER 11

Integration Theorems

A centerpiece of the calculus is the "Fundamental Theorem." This important result says that when you take a function, differentiate it, and then integrate the result something special happens. In this chapter, we explore what happens when we do the same with vector fields. As there are many different ways to both differentiate and integrate a vector field this gives us many different theorems. It should be pointed out that all of these results (including the Fundamental Theorem of Calculus) are special cases of the *generalized Stokes' Theorem*, which we will not cover here. This approach uses the theory of differential forms, rather than vector calculus, and can be found in the book *A Geometric Approach to Differential Forms*, published by Birkhäuser, 2006.

11.1 Path Independence

The first way we saw vector fields related to derivatives was through the gradient. In this section we see what happens when you start with a function f, take its gradient to get a vector field, and then perform a line integral along a curve C on this field.

Formally, we study integrals of the form

$$\int_C \nabla f \cdot d\mathbf{s}$$

To really understand what is going on we need to go back to the definition of a line integral. Suppose C is parameterized by $\Psi(t)$, and the domain of Ψ is the interval $[a, b]$. The steps to define the integral of ∇f over C are then

1. Choose n evenly spaced points in the interval $[a, b]$. Call these points $\{t_i\}$. This gives us n point $\{\Psi(t_i\}$ on the curve C.
2. Let $\Delta t = t_{i+1} - t_i$.
3. For each i compute $\nabla f \cdot \frac{d\Psi}{dt} \Delta t$ at the point $\Psi(t_i)$.
4. Sum over all i.
5. The integral $\int_C \nabla f \cdot d\mathbf{s}$ is defined to be the limit of the sum from the previous step, as $n \to \infty$.

Let us now focus on the quantity computed in Step 3 above. To estimate $\nabla f \cdot V$ at a point p we may substitute

$$\nabla f \cdot V \approx \frac{f(p) - f(p + V \Delta t)}{\Delta t}$$

This approximation gets better and better as Δt approaches zero. Furthermore, if Δt is small one can also make the approximation

$$\frac{d\Psi}{dt} \approx \frac{\Psi(t_{i+1}) - \Psi(t_i)}{t_{i+1} - t_i}$$

Putting these together we have

$$\nabla f \cdot \frac{d\Psi}{dt} \approx \frac{f(\Psi(t_{i+1})) - f(\Psi(t_i))}{\Delta t}$$

Hence, the term $\nabla f \cdot \frac{d\Psi}{dt} \Delta t$ computed in Step 3 above can be approximated by $f(\Psi(t_{i+1})) - f(\Psi(t_i))$.

We now shift our attention to Step 4:

$$\sum_i \nabla f \cdot \frac{d\Psi}{dt} \Delta t_i \approx \sum_i f(\Psi(t_{i+1})) - f(\Psi(t_i))$$

But this last expression telescopes:

$$\sum_i f(\Psi(t_{i+1})) - f(\Psi(t_i)) = f(\Psi(t_n)) - f(\Psi(t_0)) = f(\Psi(b)) - f(\Psi(a))$$

We conclude with the following theorem:

Suppose Ψ is a parameterization of C with domain $[a, b]$ which agrees with the orientation on C. Then for any function f we have

$$\int_C \nabla f \cdot d\mathbf{s} = f(\Psi(b)) - f(\Psi(a))$$

This should look a lot to you like the Fundamental Theorem of Calculus. There is no such convenient name for it. Often it is referred to as the *path independence of line integrals of gradient fields*. This is because it says that the result of a line integral of a gradient field only depend on the endpoints of the curve, and not the path used to get from one endpoint to the other.

EXAMPLE 11-1

Suppose $f(x, y) = x^3 + y^2$. Let C be the top half of the unit circle, oriented counterclockwise. We compute $\int_C \nabla f \cdot d\mathbf{s}$.

All we really have to know is the endpoints of the curve C. The first is $(1, 0)$ and the second is $(-1, 0)$. (Which one is which is determined by the orientation on C.) Hence,

$$\int_C \nabla f \cdot d\mathbf{s} = f(-1, 0) - f(1, 0) = (-1) - (1) = -2$$

There is an important application of the independence of path of line integrals of gradient fields that may be familiar. Recall that a line integral of a vector field says something about how much *work* you have to do to move an object along a curve C in the presence of a force $\mathbf{W}$. Let's say you want to know how much work you have to do against the force of gravity to get a heavy package up a mountain. Suppose the mountain is represented by the graph of a function $f(x, y)$. That is, at the map cooridnates (x, y) the function $f(x, y)$ gives you your altitude. Then the force of gravity at the point (x, y) is proportional to the vector $\nabla f(x, y)$ (i.e., the steeper the mountain, the harder you have to work to get up it). Let's suppose you have identified your route on a map. You start at the point (x_0, y_0), follow some curve C, and end up at (x_1, y_1). Then the work you have to do to overcome gravity is

$$\int_C \nabla f \cdot d\mathbf{s} = f(x_1, y_1) - f(x_0, y_0)$$

Notice that the result is just the difference in elevation between your beginning and ending point, and doesn't matter what path you take to get from one to the other!

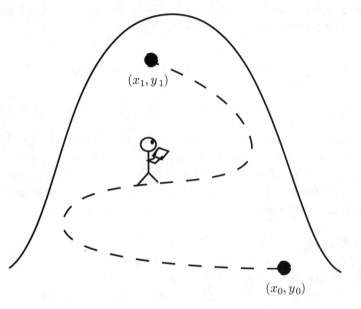

EXAMPLE 11-2

Let $\mathbf{W}$ be the vector field $\langle y^2 z^3, 2xyz^3, 3xy^2z^2 \rangle$. We would like to integrate $\mathbf{W}$ over the curve C parameterized by

$$\Psi(t) = (t^3, t^2, t), \quad 1 \le t \le 2$$

with the induced orientation.

This integral can be done directly, but the astute reader will notice that

$$\mathbf{W} = \nabla(xy^2z^3)$$

Hence, we may use the independence of path of line integrals:

$$\int_C \nabla f \cdot d\mathbf{s} = f(\Psi(2)) - f(\Psi(1)) = f(8, 4, 2) - f(1, 1, 1) = 1024 - 1 = 1023$$

Problem 116 *Let* $\mathbf{W} = \langle 1, 1, 1 \rangle$. *Suppose C is some curve that goes from* $(1, 0, 1)$ *to* $(1, 1, 1)$. *Calculate the integral of* $\mathbf{W}$ *over C.*

Problem 117 *Let C be the curve pictured below. Let* $\mathbf{W} = \langle \sin y, x \cos y \rangle$. *Calculate* $\int_C \mathbf{W} \cdot d\mathbf{s}$.

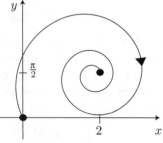

Problem 118 *Suppose* $f(x, y) = x^2 - 4x + 4 + y^2 + 2y + 1$. *Let C be a curve that starts at* $(2, -1)$ *and ends at some other point. Show that* $\int_C \nabla f \cdot d\mathbf{s}$ *is (strictly) larger than zero.*

Problem 119 *Suppose C is a closed curve, i.e., one whose beginning and ending points are the same. Show that the integral of any gradient field over C is zero.*

Problem 120 *Deduce the Fundamental Theorem of Calculus from the independence of path of line integrals of gradient fields. (Hint: Begin by letting C be a curve in* $\mathbb{R}^1$ *from a to b.)*

Problem 121 *In this section we saw that the work you must do to carry a package halfway up a mountain is just proportional to the difference in your starting and ending altitudes, and does not depend on the path you take. But it certainly seems like if you started at the bottom of a mountain, went to the top, and then came down to the halfway point, you'd be doing more work than if you just went straight up to the halfway point. Explain.*

11.2 Green's Theorem on Rectangular Domains

The next theorem that relates derivatives, integrals, and vector fields is special to $\mathbb{R}^2$. Recall that the curl of a vector field

$$\mathbf{V} = \langle f(x, y, z), g(x, y, z), h(x, y, z) \rangle$$

in $\mathbb{R}^3$ is defined to be the quantity

$$\nabla \times \mathbf{V} = \begin{vmatrix} \mathbf{i} & \mathbf{j} & \mathbf{k} \\ \dfrac{\partial}{\partial x} & \dfrac{\partial}{\partial y} & \dfrac{\partial}{\partial z} \\ f & g & h \end{vmatrix}$$

Although the curl is only defined for vector fields in $\mathbb{R}^3$, we can use it to define a natural (and important!) operation on vector fields in $\mathbb{R}^2$. First, given a vector field $\mathbf{W} = \langle f(x, y), g(x, y)\rangle$ on $\mathbb{R}^2$ we can easily create a vector field on $\mathbb{R}^3$: $\langle f(x, y), g(x, y), 0\rangle$. We will abuse the notation and continue to call this new vector field $\mathbf{W}$. We now compute its curl:

$$\nabla \times \mathbf{W} = \begin{vmatrix} \mathbf{i} & \mathbf{j} & \mathbf{k} \\ \dfrac{\partial}{\partial x} & \dfrac{\partial}{\partial y} & \dfrac{\partial}{\partial z} \\ f(x, y) & g(x, y) & 0 \end{vmatrix}$$

$$= \left(-\frac{\partial g}{\partial z}\right)\mathbf{i} - \left(-\frac{\partial f}{\partial z}\right)\mathbf{j} + \left(\frac{\partial g}{\partial x} - \frac{\partial f}{\partial y}\right)\mathbf{k}$$

$$= \left(\frac{\partial g}{\partial x} - \frac{\partial f}{\partial y}\right)\mathbf{k}$$

Hence,

$$|\nabla \times \mathbf{W}| = \frac{\partial g}{\partial x} - \frac{\partial f}{\partial y}$$

So, if $\mathbf{W}$ is a vector field on $\mathbb{R}^2$ then the result of $|\nabla \times \mathbf{W}|$ is a function on $\mathbb{R}^2$. This function acts much like the derivative of $\mathbf{W}$. We now ask, what happens when we integrate the result of such a derivative? That is, if Q is some domain in $\mathbb{R}^2$, then what can we say about

$$\iint_Q |\nabla \times \mathbf{W}| \, dx \, dy = \iint_Q \frac{\partial g}{\partial x} - \frac{\partial f}{\partial y} \, dx \, dy$$

For simplicity, we will assume here that the domain of integration Q is a rectangle. In the next section, we will explore more general domains.

To understand what is going on here, we will have to go all the way back to the definition of a multiple integral. Recall the steps to define the value of $\iint_Q h(x, y) \, dx \, dy$:

1. Choose a lattice of evenly spaced points (x_i, y_j) in Q.
2. Define

$$\Delta x = x_{i+1} - x_i, \quad \Delta y = y_{j+1} - y_j$$

3. For each i and j compute $h(x_i, y_j)\Delta x \Delta y$.
4. Sum over all i and j.
5. Take the limits of the resulting number as Δx and Δy approach zero.

To define the value of $\iint\limits_{Q} \frac{\partial g}{\partial x} - \frac{\partial f}{\partial y} \, dx \, dy$ we follow the above steps, with $h(x, y) = \frac{\partial g}{\partial x} - \frac{\partial f}{\partial y}$. Hence, in Step 3 above we compute

$$\left(\frac{\partial g}{\partial x} - \frac{\partial f}{\partial y} \right)(x_i, y_j)\Delta x \Delta y$$

If Δx and Δy are small, we may make the following approximations:

$$\frac{\partial g}{\partial x}(x_i, y_j) \approx \frac{g(x_{i+1}, y_j) - g(x_i, y_j)}{\Delta x}$$

and

$$\frac{\partial f}{\partial y}(x_i, y_j) \approx \frac{f(x_i, y_{j+1}) - f(x_i, y_j)}{\Delta y}$$

Hence, the quantity computed in Step 3 can be approximated by

$$\left(\frac{\partial g}{\partial x} - \frac{\partial f}{\partial y} \right)(x_i, y_j)\Delta x \Delta y$$

$$\approx \left(\frac{g(x_{i+1}, y_j) - g(x_i, y_j)}{\Delta x} - \frac{f(x_i, y_{j+1}) - f(x_i, y_j)}{\Delta y} \right) \Delta x \Delta y$$

$$= \left(g(x_{i+1}, y_j) - g(x_i, y_j) \right) \Delta y - \left(f(x_i, y_{j+1}) - f(x_i, y_j) \right) \Delta x$$

We now assume Q is the rectangle pictured in Figure 11-1. As in the figure, let L, R, B, and T denote the Left, Right, Bottom, and Top sides of this rectangle, with the indicated orientations.

We now move on to Step 4. In this step, we are instructed to compute the sum of the values in Step 3. Using our approximation of each term, we can approximate

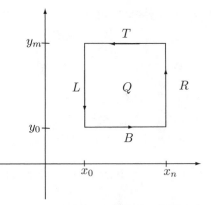

Figure 11-1   The rectangle, Q

this sum by

$$\sum_{i=0}^{n}\sum_{j=0}^{m}(g(x_{i+1}, y_j) - g(x_i, y_j))\Delta y - (f(x_i, y_{j+1}) - f(x_i, y_j))\Delta x$$

$$= \sum_{j=0}^{m}\sum_{i=0}^{n}(g(x_{i+1}, y_j) - g(x_i, y_j))\Delta y - \sum_{i=0}^{n}\sum_{j=0}^{m}(f(x_i, y_{j+1}) - f(x_i, y_j))\Delta x$$

$$= \sum_{j=0}^{m}(g(x_n, y_j) - g(x_0, y_j))\Delta y - \sum_{i=0}^{n}(f(x_i, y_m) - f(x_i, y_0))\,\Delta x$$

$$= \sum_{j=0}^{m}g(x_n, y_j)\Delta y - \sum_{j=0}^{m}g(x_0, y_j)\Delta y - \sum_{i=0}^{n}f(x_i, y_m)\Delta x + \sum_{i=0}^{n}f(x_i, y_0)\Delta x$$

Note that each term in the final expression is a sum involving values of f or g at points on the sides of Q. For example, in the first sum we see the expression $g(x_n, y_j)$. The number n is a constant, and represents the largest index of the x-values. Hence, a point of the form (x_n, y_j) must lie on R, the Right side of Q.

We are now prepared to consider Step 5, where we take limits as Δx and Δy go to zero. We do this for each term in the above sum. For example, consider the limit of the first term:

$$\lim_{\Delta x, \Delta y \to 0}\sum_{j=0}^{m}g(x_n, y_j)\Delta y = \lim_{\Delta y \to 0}\sum_{j=0}^{m}g(x_n, y_j)\Delta y$$

$$\int_{y_0}^{y_m} g(x_n, y)\,dy$$

But this integral is the same as the integral of **W** over R with the orientation indicated in Figure 11-1. To see this, first parameterize R by

$$\Psi(y) = (x_n, y), \quad y_0 \le y \le y_m$$

Then,

$$\frac{d\Psi}{dy} = \langle 0, 1 \rangle$$

and so

$$\int_R \mathbf{W} \cdot d\mathbf{s} = \int_{y_0}^{y_m} \langle f(x_n, y), g(x_n, y) \rangle \cdot \langle 0, 1 \rangle dy$$

$$= \int_{y_0}^{y_m} g(x_n, y) \, dy$$

The other three terms in the sum from Step 4 give similar integrals over the remaining sides of Q. We are now ready for the conclusion:

$$\iint_Q \frac{\partial g}{\partial x} - \frac{\partial f}{\partial y} \, dx \, dy = \int_R \mathbf{W} \cdot d\mathbf{s} + \int_L \mathbf{W} \cdot d\mathbf{s} + \int_T \mathbf{W} \cdot d\mathbf{s} + \int_B \mathbf{W} \cdot d\mathbf{s}$$

The sides R, L, T, and B of Q, with the orientations as indicated in Figure 11-1, when taken together are referred to as its *boundary*. The usual notation for this is ∂Q. Using this notation we can write our conclusion much more succinctly:

$$\iint_Q \frac{\partial g}{\partial x} - \frac{\partial f}{\partial y} \, dx \, dy = \int_{\partial Q} \mathbf{W} \cdot d\mathbf{s}$$

This final equation is known as "Green's Theorem."

EXAMPLE 11-3
Let Q be the rectangle in the plane with corners at $(0, 0)$, $(2, 0)$, $(0, 3)$, and $(2, 3)$. Let $\mathbf{W} = \langle xy, x^2 + y^2 \rangle$. We will use Green's Theorem to evaluate the integral of

$\mathbf{W}$ over ∂Q.

$$\int_{\partial Q} \mathbf{W} \cdot d\mathbf{s} = \iint_Q 2x - x \, dx \, dy$$

$$= \int_0^3 \int_0^2 x \, dx \, dy$$

$$= \int_0^3 2 \, dy$$

$$= 6$$

Green's Theorem enables us to see the geometric significance of the value of $\frac{\partial g}{\partial x} - \frac{\partial f}{\partial y}$ at a point (x_0, y_0). Let Q denote a *very* small rectangle around this point. Then at each point of Q the value of $\frac{\partial g}{\partial x} - \frac{\partial f}{\partial y}$ is roughly constant, and equal to its value at (x_0, y_0). Hence,

$$\iint_Q \frac{\partial g}{\partial x} - \frac{\partial f}{\partial y} \, dx \, dy \approx \left(\frac{\partial g}{\partial x}(x_0, y_0) - \frac{\partial f}{\partial y}(x_0, y_0) \right) \iint_Q dx \, dy$$

$$= \left(\frac{\partial g}{\partial x}(x_0, y_0) - \frac{\partial f}{\partial y}(x_0, y_0) \right) \text{Area}(Q)$$

But Green's Theorem says

$$\iint_Q \frac{\partial g}{\partial x} - \frac{\partial f}{\partial y} \, dx \, dy = \int_{\partial Q} \mathbf{W} \cdot d\mathbf{s}$$

Putting these together gives us

$$\left(\frac{\partial g}{\partial x}(x_0, y_0) - \frac{\partial f}{\partial y}(x_0, y_0) \right) \text{Area}(Q) \approx \int_{\partial Q} \mathbf{W} \cdot d\mathbf{s}$$

or,

$$\frac{\partial g}{\partial x}(x_0, y_0) - \frac{\partial f}{\partial y}(x_0, y_0) \approx \frac{1}{\text{Area}(Q)} \int_{\partial Q} \mathbf{W} \cdot d\mathbf{s}$$

So, the function $\frac{\partial g}{\partial x} - \frac{\partial f}{\partial y}$ is a measure of the "circulation" of $\mathbf{W}$ around each point. There is a way to experience such a function for yourself. Many people enjoy going "tubing"—floating down a river in an inner tube. Suppose you are tubing and decide you want to stop somewhere to enjoy the scenery, so you drop an anchor. Then you notice that the water on your left is rushing past you faster than the water on your right. What happens? Your inner tube starts to turn. This turning is a measure of the strength of $\frac{\partial g}{\partial x} - \frac{\partial f}{\partial y}$.

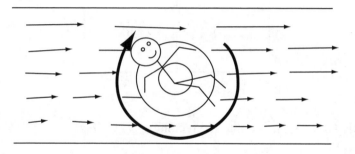

Problem 122 *Let Q be the rectangle $\{(x, y)|0 \le x \le 1, 0 \le y \le 1\}$. Use Green's Theorem to evaluate the integral of $\langle -y^2, x^2 \rangle$ over ∂Q.*

Problem 123 *Let C be the curve pictured below. Show that the integral of $\mathbf{W} = \langle y, x \rangle$ over C does not depend on b.*

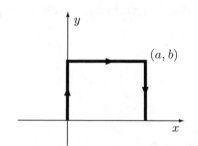

11.3 Green's Theorem over More General Domains

Although we have only demonstrated the validity of Green's Theorem for rectangular regions, it holds in much more generality. For other shaped regions Q the expression ∂Q should generally be interpreted as the "edge" of Q, with a counterclockwise orientation. But there is an important technical restriction on the types of regions Q for which this is valid.

We begin by examining what happens when we look at Green's Theorem applied to two neighboring rectangles.

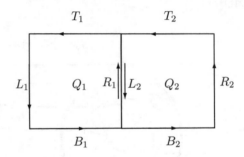

We apply Green's Theorem to the rectangle $Q_1 \cup Q_2$:

$$\iint\limits_{Q_1 \cup Q_2} \frac{\partial g}{\partial x} - \frac{\partial f}{\partial y}\, dx\, dy = \iint\limits_{Q_1} \frac{\partial g}{\partial x} - \frac{\partial f}{\partial y}\, dx\, dy + \iint\limits_{Q_2} \frac{\partial g}{\partial x} - \frac{\partial f}{\partial y}\, dx\, dy$$

$$= \int\limits_{\partial Q_1} \mathbf{W} \cdot d\mathbf{s} + \int\limits_{\partial Q_2} \mathbf{W} \cdot d\mathbf{s}$$

But the integral of $\mathbf{W}$ along R_1 will cancel with the integral of $\mathbf{W}$ along L_2, so the last integral equals

$$\int\limits_{\partial (Q_1 \cup Q_2)} \mathbf{W} \cdot d\mathbf{s}$$

as Green's Theorem would predict.

The cancellation of the integrals over R_1 and L_2 illustrates an important phenomenon. For example, suppose Q is the following region.

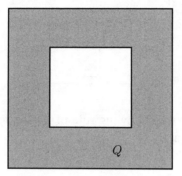

We can use Green's Theorem to analyze the integral of $\frac{\partial g}{\partial x} - \frac{\partial f}{\partial y}$ over Q by breaking it up into rectangles. The boundary of each rectangle then gets an orientation. When we add up the integral of **W** over the boundary of every rectangle, there is much cancellation, as in the following figure.

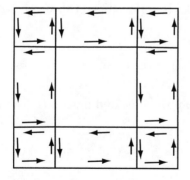

We end up with the integral of $\frac{\partial g}{\partial x} - \frac{\partial f}{\partial y}$ being equal to the integral of **W** over the squares depicted in the following figure, with the indicated orientations. Hence, we say that this is the appropriate boundary of Q.

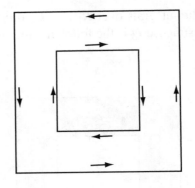

In general, we may use the following rule-of-thumb to figure out the orientation on each loop of its boundary:

> If Q is a connected region in $\mathbb{R}^2$ then the "outermost" loop of its boundary is oriented counterclockwise and all other loops of the boundary are oriented clockwise.

EXAMPLE 11-4

We use Green's Theorem to integrate the function $x^2 + y^2$ over the inside of the unit circle. If we denote this region as Q, then the boundary of Q is the unit circle itself, with a counterclockwise orientation. This can be parameterized in the usual way by

$$\Psi(t) = (\cos t, \sin t), \quad 0 \le t \le 2\pi$$

To use Green's Theorem we must find functions f and g so that

$$\frac{\partial g}{\partial x} - \frac{\partial f}{\partial y} = x^2 + y^2 = y^2 - (-x^2)$$

A suitable choice for $g(x, y)$ can be found by integrating y^2 with respect to x, yielding the function xy^2. Similarly, we may find $f(x, y)$ by integrating $-x^2$ with respect to y, yielding $-x^2 y$. So

$$\langle f(x, y), g(x, y) \rangle = \langle -x^2 y, xy^2 \rangle$$

Finally, we integrate

$$\iint_Q x^2 + y^2 \, dx \, dy = \int_{\partial Q} \langle -x^2 y, xy^2 \rangle \cdot d\mathbf{s}$$

$$= \int_0^{2\pi} \langle -\cos^2 t \sin t, \cos t \sin^2 t \rangle \cdot \frac{d\Psi}{dt} \, dt$$

$$= \int_0^{2\pi} \langle -\cos^2 t \sin t, \cos t \sin^2 t \rangle \cdot \langle -\sin t, \cos t \rangle \, dt$$

$$= \int_0^{2\pi} \cos^2 t \sin^2 t + \cos^2 t \sin^2 t \, dt$$

$$= \int_0^{2\pi} 2 \cos^2 t \sin^2 t \, dt$$

$$= \int_0^{2\pi} \frac{1}{2} \sin^2 2t \, dt$$

$$= \int_0^{4\pi} \frac{1}{4} \sin^2 u \, du$$

$$= \frac{1}{4} \left(\frac{1}{2}u - \frac{1}{4} \sin 2u \right) \Big|_0^{4\pi}$$

$$= \frac{\pi}{2}$$

Problem 124 *Let* $\mathbf{W} = \langle -y^2, x^2 \rangle$. *Let* σ *be the region in* $\mathbb{R}^2$ *parameterized by the following:*

$$\phi(u, v) = (2u - v, u + v)$$

where $1 \leq u \leq 2$ *and* $0 \leq v \leq 1$. *Calculate* $\int_{\partial \sigma} \mathbf{W} \cdot d\mathbf{s}$.

Problem 125 *Calculate the area enclosed by the unit circle by integrating some vector field around its boundary.*

Problem 126

1. *Suppose* $\mathbf{W} = \langle f(x), g(x) \rangle$ *is a vector field which is defined everywhere except at* $(0,0)$. *If* $\frac{\partial g}{\partial x} - \frac{\partial f}{\partial y} = 0$ *then show that the integral of* $\mathbf{W}$ *along every circle centered on the origin, oriented counterclockwise, is the same.*
2. *If* $\mathbf{W} = \langle \frac{-y}{x^2+y^2}, \frac{x}{x^2+y^2} \rangle$ *then show that the integral of* $\mathbf{W}$ *along every circle centered on the origin, oriented counterclockwise, is the same.*
3. *Calculate the integral of* $\langle \frac{-y}{x^2+y^2}, \frac{x}{x^2+y^2} \rangle$ *over the unit circle.*

Problem 127 *Let* σ *be the region parameterized by*

$$\phi(r, \theta) = (r\cos\theta, r\sin\theta), \ 0 \le r \le 1, \ 0 \le \theta \le \pi$$

Suppose $\mathbf{W} = \langle x^2, e^y \rangle$.

1. *Use Green's Theorem to show that* $\int_{\partial\sigma} \mathbf{W} \cdot d\mathbf{s} = 0$.
2. *Let* C *be the horizontal segment connecting* $(-1, 0)$ *to* $(1, 0)$. *Calculate* $\int_C \mathbf{W} \cdot d\mathbf{s}$.
3. *Use your previous answers to determine the integral of* $\mathbf{W}$ *over the top half of the unit circle (oriented counterclockwise).*

11.4 Stokes' Theorem

In the previous section, we saw that if $\mathbf{W}$ is a vector field in $\mathbb{R}^2$, then we can view $|\nabla \times \mathbf{W}|$ as a kind of derivative. When we integrated this "derivative" we saw something special happen, namely Green's Theorem:

$$\iint_Q |\nabla \times \mathbf{W}| \ dx \ dy = \int_{\partial Q} \mathbf{W} \cdot d\mathbf{s}$$

We now move our attention to $\mathbb{R}^3$, and explore a similar phenomenon. Suppose now $\mathbf{W}$ is a vector field in $\mathbb{R}^3$, and S is a surface. Then we wish to explore $\int_S (\nabla \times \mathbf{W}) \cdot d\mathbf{S}$. A reasonable guess, based on our experience from the previous section, would be

$$\int_S (\nabla \times \mathbf{W}) \cdot d\mathbf{S} = \int_{\partial S} \mathbf{W} \cdot d\mathbf{s}$$

This turns out to be the case, and is called *Stokes' Theorem*. We will not prove it here, as the proof is extremely similar to that of Green's Theorem. The strategy, once again, is to choose a lattice of points in S. This breaks up S into a bunch of parallelograms, and we can chase through the definition of $\int_S (\nabla \times \mathbf{W}) \cdot d\mathbf{S}$ on each. At every parallelogram, we see that we get the same as the integral of $\mathbf{W}$ over the boundary of the parallelogram. But, because of orientation considerations, the integrals over the boundaries of neighboring parallelograms cancel. The result is the integral of $\mathbf{W}$ over the boundary of S.

One potential complication in using Stokes' Theorem is determining the boundary of the surface S in question. To get the proper orientation on ∂S you need to know the orientation of S. Recall that this is often given by an outward-pointing normal vector, v. To get the orientation on the boundary, we use the "right-hand rule." To do this point the thumb of your right hand in the direction of v. Your fingers will then curl in the sense that determines the orientation on the boundary.

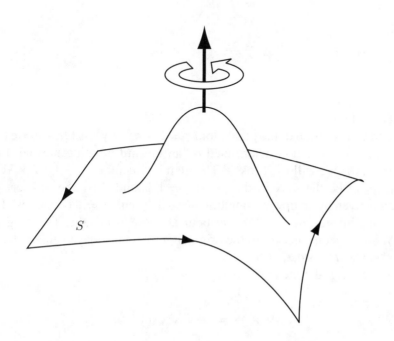

EXAMPLE 11-5

Let S denote the top half of the unit sphere, with orientation given by the normal vector $\langle 1, 0, 0 \rangle$ at the point $(1, 0, 0)$. We use Stokes' Theorem to integrate the curl of the vector field $\langle -y, x, 0 \rangle$ over S.

First, note that Stokes' Theorem says that the answer will be the same as the integral of $\langle -y, x, 0 \rangle$ around ∂S. The boundary of S (with proper orientation) is parameterized by

$$\Psi(t) = (\cos t, \sin t, 0), \quad 0 \leq t \leq 2\pi$$

Thus, we may integrate

$$\int\limits_S (\nabla \times \langle -y, x \rangle) \cdot d\mathbf{S} = \int\limits_{\partial S} \langle -y, x, 0 \rangle \cdot d\mathbf{s}$$

$$= \int\limits_0^{2\pi} \langle -\sin t, \cos t, 0 \rangle \cdot \langle -\sin t, \cos t, 0 \rangle \, dt$$

$$= \int\limits_0^{2\pi} dt$$

$$= 2\pi$$

EXAMPLE 11-6

Let S denote the portion of the paraboloid $z = 2 - x^2 - y^2$ that lies above the plane $z = 1$, with an orientation determined by an upward pointing normal. Let $\mathbf{W} = \langle \cos z, \sin z, 0 \rangle$. We will use Stokes' Theorem indirectly to find $\int_S (\nabla \times \mathbf{W}) \cdot d\mathbf{S}$.

First, let D be the disk in the plane $z = 1$ bounded by the unit circle, with orientation given by an upward pointing normal. Then $\partial S = \partial D$. Stokes' Theorem says that the integral of $\nabla \times \mathbf{W}$ over both D and S is equal to the integral of $\mathbf{W}$ over ∂S. So, to get an answer to the original problem we may evaluate the integral of $\nabla \times \mathbf{W}$ over D instead of S.

To do the integral, note that

$$\nabla \times \mathbf{W} = \langle -\cos z, \sin z, 0 \rangle$$

So, on the plane $z = 4$ we have $\nabla \times \mathbf{W} = \langle -\cos 1, \sin 1, 0 \rangle$. A parameterization for D is given by

$$\Psi(r, \theta) = (r \cos \theta, r \sin \theta, 1), \quad 0 \le r \le 1, \quad 0 \le \theta \le 2\pi$$

The reader may check that

$$\frac{\partial \Psi}{\partial r} \times \frac{\partial \Psi}{\partial \theta} = \langle 0, 0, r \rangle$$

We now integrate

$$\int_D (\nabla \times \mathbf{W}) \cdot d\mathbf{S} = \int_0^1 \int_0^{2\pi} \langle -\cos 1, \sin 1, 0 \rangle \cdot \langle 0, 0, r \rangle \, dr \, d\theta$$
$$= 0$$

Problem 128 *Let* $\mathbf{W} = \langle xy, xz, y \rangle$. *Let S be the surface parameterized by*

$$\Psi(r, \theta) = (0, r \cos \theta, r \sin \theta), \quad 0 \le r \le 1, \quad 0 \le \theta \le 2\pi$$

with the induced orientation. Calculate $\int_S (\nabla \times \mathbf{W}) \cdot d\mathbf{S}$.

Problem 129 *Suppose* $\mathbf{W} = \langle f(x, y), g(x, y), 0 \rangle$ *and S is a region of* $\mathbb{R}^3$ *that lies in the xy-plane. Show that Stokes' Theorem applied to* $\mathbf{W}$ *and S is equivalent to Green's Theorem.*

Problem 130 *Let S be the portion of the cylinder* $x^2 + y^2 = 1$ *that lies between the planes* $z = 0$ *and* $z = 1$, *with orientation given by the normal vector* $\langle 1, 0, 0 \rangle$ *at the point* $(1, 0, 0)$. *Let* $\mathbf{W} = \langle -yz, xz, 0 \rangle$. *Calculate the integral of* $\nabla \times \mathbf{W}$ *over S.*

Problem 131 *If* $\mathbf{W}$ *is a vector field defined on all of* $\mathbb{R}^3$, *then show that the integral of* $\nabla \times \mathbf{W}$ *over the unit sphere is zero.*

11.5 Geometric Interpretation of Curl

Just as Green's Theorem gave us a way to interpret the function $\frac{\partial g}{\partial x} - \frac{\partial f}{\partial y}$ geometrically, we can use Stokes' Theorem to give a geometric interpretation of the curl of a vector field. Let $\mathbf{W} = \langle f, g, h \rangle$ be a vector field on $\mathbb{R}^3$ and p a point of $\mathbb{R}^3$. Our goal is to understand the meaning of the vector $\nabla \times \mathbf{W}(p)$.

Let D be a small, flat disk centered on p. If D is small enough, then $\nabla \times \mathbf{W}$ is roughly constant at every point of D. Let $\Psi(u, v)$ denote a parameterization of D, with domain R, in which

$$\left| \frac{\partial \Psi}{\partial u} \times \frac{\partial \Psi}{\partial v} \right| = 1$$

In particular, this implies

$$\text{Area}(D) = \iint_R \left| \frac{\partial \Psi}{\partial u} \times \frac{\partial \Psi}{\partial v} \right| \, du \, dv$$

$$= \iint_R du \, dv$$

$$= \text{Area}(R)$$

Since D is part of a plane the vectors normal to D are all parallel. Hence, our assumption that the magnitude of $\frac{\partial \Psi}{\partial u} \times \frac{\partial \Psi}{\partial v}$ is constant implies

$$\frac{\partial \Psi}{\partial u} \times \frac{\partial \Psi}{\partial v} = N$$

for some fixed unit vector N which is normal to D.

We now examine the integral of the curl of $\mathbf{W}$ over D:

$$\int_D (\nabla \times \mathbf{W}) \cdot d\mathbf{S} = \iint_R \nabla \times \mathbf{W}(\Psi(u, v)) \cdot \left(\frac{\partial \Psi}{\partial u} \times \frac{\partial \Psi}{\partial v} \right) \, du \, dv$$

$$= \iint_R \nabla \times \mathbf{W}(\Psi(u, v)) \cdot N \, du \, dv$$

$$\approx (\nabla \times \mathbf{W}(p)) \cdot N \iint_R du \, dv$$

$$= (\nabla \times \mathbf{W}(p)) \cdot N \operatorname{Area}(R)$$

$$= (\nabla \times \mathbf{W}(p)) \cdot N \operatorname{Area}(D)$$

Recall that the dot product of two vectors is the product of their magnitudes times the cosine of the angle between them. Hence, if we choose D so that N points in the same direction as $\nabla \times \mathbf{W}(p)$ we get

$$(\nabla \times \mathbf{W}(p)) \cdot N = |\nabla \times \mathbf{W}(p)|$$

and hence

$$\int_{D} (\nabla \times \mathbf{W}) \cdot d\mathbf{S} \approx |\nabla \times \mathbf{W}(p)| \operatorname{Area}(D)$$

To go further we must appeal to Stokes' Theorem:

$$\int_{D} (\nabla \times \mathbf{W}) \cdot d\mathbf{S} = \int_{\partial D} \mathbf{W} \cdot d\mathbf{s}$$

Putting these together then gives

$$\int_{\partial D} \mathbf{W} \cdot d\mathbf{s} \approx |\nabla \times \mathbf{W}(p)| \operatorname{Area}(D)$$

or,

$$|\nabla \times \mathbf{W}(p)| \approx \frac{1}{\operatorname{Area}(D)} \int_{\partial D} \mathbf{W} \cdot d\mathbf{s}$$

Our conclusion is that the magnitude of $\nabla \times \mathbf{W}$ at the point p is a measure of how much $\mathbf{W}$ circulates around it. The direction of $\nabla \times \mathbf{W}$ is the same as the direction of N, which was chosen to be perpendicular to the plane in which this circulation is greatest. In this sense, the curl of a vector field is the three-dimensional version of the function $\frac{\partial g}{\partial x} - \frac{\partial f}{\partial y}$, which appears in Green's Theorem.

EXAMPLE 11-7

Let $\mathbf{W} = \langle -y, x, 0 \rangle$. Then

$$\nabla \times \mathbf{W} = \begin{vmatrix} \mathbf{i} & \mathbf{j} & \mathbf{k} \\ \dfrac{\partial}{\partial x} & \dfrac{\partial}{\partial y} & \dfrac{\partial}{\partial z} \\ -y & x & 0 \end{vmatrix} = \langle 0, 0, 2 \rangle$$

This is a vector that points up. Notice that a plane perpendicular to an upward pointing vector is the plane that contains the "circulation" of $\mathbf{W}$. In contrast, the circulation in a plane which contains $\nabla \times \mathbf{W}$ would be zero.

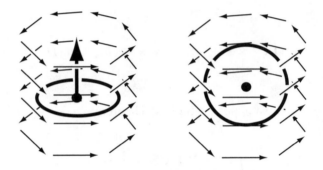

Problem 132 *Among all circles C in $\mathbb{R}^3$ centered at the origin with radius 1, suppose the circulation $\int_C \mathbf{W} \cdot d\mathbf{s}$ is greatest when C lies in the xz-plane. Suppose, furthermore, that when C is such a loop,*

$$\int_C \mathbf{W} \cdot d\mathbf{s} = .5$$

Estimate $\nabla \times \mathbf{W}$ at the origin.

Problem 133 *Suppose some vector field has the property that the direction of every vector is up, and in any vertical plane parallel to the xz-plane the vector field is constant. Then what direction would the curl of this vector field point?*

11.6 Gauss' Theorem

We know that one way to "differentiate" a vector field is to take its curl. It is then not surprising that the integral of the curl of a vector field should be special. Another way to "differentiate" a vector field in $\mathbb{R}^3$ is to take its divergence. In this section, we explore what happens when we integrate the divergence of a vector field. To this end, suppose

$$\mathbf{W} = \langle f(x, y, z), g(x, y, z), h(x, y, z) \rangle$$

Recall the definition of divergence:

$$\text{Div } \mathbf{W} = \nabla \cdot \mathbf{W} = \frac{\partial f}{\partial x} + \frac{\partial g}{\partial y} + \frac{\partial h}{\partial z}$$

The result is a function on $\mathbb{R}^3$. We may thus integrate this function over volumes V:

$$\iiint\limits_{V} \nabla \cdot \mathbf{W} \, dx \, dy \, dz$$

As in the previous sections, we might guess that there is a relationship between this and the integral of $\mathbf{W}$ over the boundary of V:

$$\iiint\limits_{V} \nabla \cdot \mathbf{W} \, dx \, dy \, dz = \int\limits_{\partial V} \mathbf{W} \cdot d\mathbf{S}$$

This equality is in fact true, and is known as *Gauss' Divergence Theorem*. The proof is again similar to the proof of Green's Theorem. We choose a three-dimensional lattice of points in V, and approximate $\iiint\limits_{V} \nabla \cdot \mathbf{W} \, dx \, dy \, dz$ as a sum over the points of this lattice. The lattice breaks up V into little cubes, and we find that the integral of $\nabla \cdot \mathbf{W}$ over each cube is approximately equal to the integral of $\mathbf{W}$ over the boundary of each cube, with suitable orientations. But faces of cubes inherit opposite orientations from neighboring cubes, so in the sum all that is left are the faces of the cubes on the boundary of V.

To properly orient ∂V, we simply choose a normal vector that points "out" of V.

EXAMPLE 11-8
Let $\mathbf{W} = \langle x, y, z \rangle$. We would like to find the value of the integral of $\nabla \cdot \mathbf{W}$ over the volume V bounded by the unit sphere. According to Gauss' Theorem, this is equal to the integral of $\mathbf{W}$ over the unit sphere S.

To evaluate this, we first parameterize the unit sphere in the usual way with spherical coordinates:

$$\Psi(\theta, \phi) = (\sin\phi\cos\theta, \sin\phi\sin\theta, \cos\phi)$$

$$0 \le \theta \le 2\pi, \quad 0 \le \phi \le \pi$$

Now we compute

$$\frac{\partial\Psi}{\partial\theta} \times \frac{\partial\Psi}{\partial\phi} = \langle -\sin\phi\sin\theta, \sin\phi\cos\theta, 0\rangle \times \langle \cos\phi\cos\theta, \cos\phi\sin\theta, -\sin\phi\rangle$$

$$= \langle -\sin^2\phi\cos\theta, -\sin^2\phi\sin\theta, -\sin\phi\cos\phi\rangle$$

To check orientations, note that at $\Psi(0, \frac{\pi}{2}) = (1, 0, 0)$ this vector is equal to $\langle 1, 0, 0\rangle$, which does indeed point out of V. Hence, we do not need to worry about negating the value of an integral that is computed using Ψ.

Finally, we integrate:

$$\iiint\limits_{V} \nabla \cdot \mathbf{W} \, dx \, dy \, dz$$

$$= \int\limits_{S} \mathbf{W} \cdot d\mathbf{S}$$

$$= \int_0^\pi \int_0^{2\pi} \langle \sin\phi\cos\theta, \sin\phi\sin\theta, \cos\phi\rangle$$

$$\times \langle -\sin^2\phi\cos\theta, -\sin^2\phi\sin\theta, -\sin\phi\cos\phi\rangle \, d\theta \, d\phi$$

$$= \int_0^\pi \int_0^{2\pi} -\sin^3\phi - \sin\phi\cos\phi \, d\theta \, d\phi$$

$$= \int_0^\pi \int_0^{2\pi} -\sin\phi \, d\theta \, d\phi$$

$$= 2\pi \int_0^\pi -\sin\phi \, d\phi$$

$$= -4\pi$$

EXAMPLE 11-9
Let $\mathbf{W} = \langle 0, 0, z^2 \rangle$. We integrate $\nabla \cdot \mathbf{W}$ over the volume V which is inside the cylinder $x^2 + y^2 = 1$, above the plane $z = 0$, and below the plane $z = 1$.

To use Gauss' Theorem we will have to parameterize each "side" of V. First, the cylinder, C:

$$\Psi(\theta, z) = (\cos\theta, \sin\theta, z), \quad 0 \le \theta \le 2\pi, \quad 0 \le z \le 1$$

Then the bottom, B:

$$\Psi_-(r, \theta) = (r\cos\theta, r\sin\theta, 0), \quad 0 \le r \le 1, \quad 0 \le \theta \le 2\pi$$

And finally the top, T:

$$\Psi_+(r, \theta) = (r\cos\theta, r\sin\theta, 1), \quad 0 \le r \le 1, \quad 0 \le \theta \le 2\pi$$

Now we must check orientations by computing normal vectors:

$$\frac{\partial\Psi}{\partial\theta} \times \frac{\partial\Psi}{\partial z} = \langle -\sin\theta, \cos\theta, 0 \rangle \times \langle 0, 0, 1 \rangle = \langle \cos\theta, \sin\theta, 0 \rangle$$

This vector points out of V.

$$\frac{\partial\Psi_-}{\partial r} \times \frac{\partial\Psi_-}{\partial\theta} = \langle \cos\theta, \sin\theta, 0 \rangle \times \langle -r\sin\theta, r\cos\theta, 0 \rangle = \langle 0, 0, r \rangle$$

This vector points "up," which is into V. We will have to remember to negate the value of any integral that is computed using Ψ_-.

$$\frac{\partial\Psi_+}{\partial r} \times \frac{\partial\Psi_+}{\partial\theta} = \langle \cos\theta, \sin\theta, 0 \rangle \times \langle -r\sin\theta, r\cos\theta, 0 \rangle = \langle 0, 0, r \rangle$$

This vector again points "up," but at the top this is pointing out of V.
Gauss' Theorem says

$$\iiint_V \nabla \cdot \mathbf{W} \, dx \, dy \, dz = \int_{\partial V} \mathbf{W} \cdot d\mathbf{S}$$

$$= \int_{C \cup B \cup T} \mathbf{W} \cdot d\mathbf{S}$$

$$= \int_C \mathbf{W} \cdot d\mathbf{S} + \int_B \mathbf{W} \cdot d\mathbf{S} + \int_T \mathbf{W} \cdot d\mathbf{S}$$

We compute each of these integrals individually:

$$\int_C \mathbf{W} \cdot d\mathbf{S} = \int_0^1 \int_0^{2\pi} \langle 0, 0, z^2 \rangle \times \langle \cos\theta, \sin\theta, 0 \rangle \, d\theta \, dz$$

$$= 0$$

$$\int_B \mathbf{W} \cdot d\mathbf{S} = -\int_0^{2\pi} \int_0^1 \langle 0, 0, 0 \rangle \times \langle 0, 0, r \rangle \, dr \, d\theta$$

$$= 0$$

$$\int_T \mathbf{W} \cdot d\mathbf{S} = \int_0^{2\pi} \int_0^1 \langle 0, 0, 1 \rangle \times \langle 0, 0, r \rangle \, dr \, d\theta$$

$$= \int_0^{2\pi} \int_0^1 r \, dr \, d\theta$$

$$= \pi$$

Hence,

$$\iiint_V \nabla \cdot \mathbf{W} \, dx \, dy \, dz = \pi$$

Problem 134 *Integrate* $\mathbf{W} = \langle x^2 yz, xy^2 z, xyz^2 \rangle$ *over the boundary of the unit cube in* $\mathbb{R}^3$.

Problem 135 *Let* $\mathbf{W} = \langle 0, 0, e^z \rangle$. *Calculate the integral of* $\nabla \cdot \mathbf{W}$ *over the ball bounded by the unit sphere.*

Problem 136 *Let* $\mathbf{W} = \langle x^3, y^3, 0 \rangle$. *Let V be the region between the cylinders of radii 1 and 2 (centered on the z-axis), in the positive octant, and below the plane* $z = 2$. *Calculate*

$$\int_{\partial V} \mathbf{W} \cdot d\mathbf{S}$$

Problem 137 *The surfaces C and D are defined by*

1. *C is the graph of the cylindrical equation $r = \cos\theta$ in $\mathbb{R}^3$, where $0 \le \theta \le \pi$ and $0 \le z \le 1$.*
2. *D is the set of points in the plane $z = 1$, which are within $\frac{1}{2}$ of a unit away from the point $(\frac{1}{2}, 0, 1)$.*

Let $\mathbf{W}$ be the vector field $\langle 0, xyz, 0 \rangle$. Calculate $\int_{C+D} \mathbf{W} \cdot d\mathbf{S}$.

Problem 138 *Suppose $\nabla \cdot \mathbf{W} = 0$, and S_1 and S_2 are oriented surfaces with the same oriented boundary. Show that*

$$\int_{S_1} \mathbf{W} \cdot d\mathbf{S} = \int_{S_2} \mathbf{W} \cdot d\mathbf{S}$$

(For simplicity you may assume that S_1 and S_2 only meet in their boundary.)

11.7 Geometric Interpretation of Divergence

Recall that Green's Theorem gave us a geometric interpretation of $\frac{\partial g}{\partial x} - \frac{\partial f}{\partial y}$ and Stokes' Theorem gave us a geometric interpretation for $\nabla \times \mathbf{W}$. We now use Gauss' Theorem to give a geometric interpretation of $\nabla \cdot \mathbf{W}$. This will be completely analogous.

Let $\mathbf{W}$ denote a vector field on $\mathbb{R}^3$. Let p be a point in $\mathbb{R}^3$ and let B be a small ball centered on p. If B is chosen small enough then the function $\nabla \cdot \mathbf{W}$ is approximately equal to $\nabla \cdot \mathbf{W}(p)$ on all of B. Hence,

$$\iiint_B \nabla \cdot \mathbf{W} \, dx\, dy\, dz \approx \nabla \cdot \mathbf{W}(p) \iiint_B dx\, dy\, dz$$

$$= \nabla \cdot \mathbf{W}(p) \text{Volume}(B)$$

Now, Gauss' Theorem says

$$\iiint_B \nabla \cdot \mathbf{W} \, dx\, dy\, dz = \int_{\partial B} \mathbf{W} \cdot d\mathbf{S}$$

Combining these we get

$$\nabla \cdot \mathbf{W}(p)\text{Volume}(B) \approx \int_{\partial B} \mathbf{W} \cdot d\mathbf{S}$$

or,

$$\nabla \cdot \mathbf{W}(p) \approx \frac{1}{\text{Volume}(B)} \int_{\partial B} \mathbf{W} \cdot d\mathbf{S}$$

The quantity $\int_{\partial B} \mathbf{W} \cdot d\mathbf{S}$ is a measure of the net amount of $\mathbf{W}$ which leaves the ball B. So, if the same amount of $\mathbf{W}$ enters and leaves B then this integral will be zero. One way to think about this is that $\int_{\partial B} \mathbf{W} \cdot d\mathbf{S}$, and hence $\nabla \cdot \mathbf{W}(p)$, is a measure of how much $\mathbf{W}$ "spreads out" at p.

EXAMPLE 11-10
The vector fields pictured below have more "leaving" each point than "entering," and so have positive divergence. For example, the second one pictured might be something like $\langle x, y, z \rangle$, whose divergence is 3. The third one might be $\langle 0, 0, z \rangle$, whose divergence is 1. Can you think of a vector field like the first one?

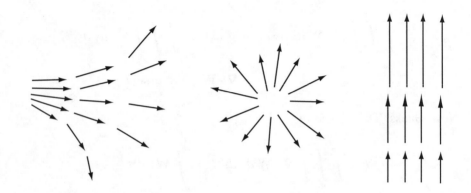

EXAMPLE 11-11

The vector fields pictured below have zero divergence. The first one is a constant vector field. The second might be something like $\langle y, -x, 0 \rangle$.

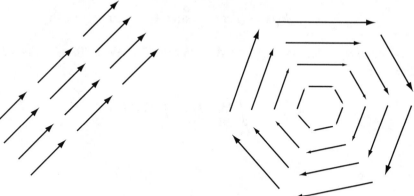

Problem 139 *Suppose B is a ball of radius 1 centered around the origin. Let* **W** *be a vector field and suppose*

$$\int \int_B \int \nabla \cdot \mathbf{W} \, dx \, dy \, dz = .5$$

Estimate the value of $\nabla \cdot \mathbf{W}$ *at the origin.*

Quiz

Problem 140

1. *Let C be any curve in* $\mathbb{R}^3$ *from* $(0, 0, 0)$ *to* $(1, 1, 1)$. *Let* **W** *be the vector field* $\langle y^2 z^2, 2xyz^2, 2xy^2 z \rangle$. *Calculate* $\int_C \mathbf{W} \cdot d\mathbf{s}$.

2. *Let* σ *be the region parameterized by the following:*

$$\phi(u, v) = (uv^2, u^3 v), \quad 1 \le u \le 2, \quad 1 \le v \le 2$$

   Calculate

$$\int_{\partial \sigma} \langle 1, -\ln x \rangle \cdot d\mathbf{s}$$

3. *Let C_1 and C_2 be curves given by the following parameterizations (with the induced orientations):*

$$C_1 : \phi(t) = (t, 0, 0), \quad 2\pi \leq t \leq 4\pi$$
$$C_2 : \psi(t) = (t \cos t, t \sin t, 0), \quad 2\pi \leq t \leq 4\pi$$

Show that for any vector field $\mathbf{W}$ such that $\nabla \times \mathbf{W} = \langle 0, 0, 0 \rangle$ the following is true:

$$\int_{C_1} \mathbf{W} \cdot d\mathbf{s} = \int_{C_2} \mathbf{W} \cdot d\mathbf{s}$$

Final Exam

Problem 141

1. Let $f(x, y)$ be the following function:

$$f(x, y) = xy + x - 2y + 4$$

   a. Sketch the intersections of $f(x, y)$ with the xz- and yz-planes.

   b. Find the critical point(s) and compute the value of

$$\begin{vmatrix} \dfrac{\partial^2 f}{\partial x^2} & \dfrac{\partial^2 f}{\partial x \partial y} \\[2mm] \dfrac{\partial^2 f}{\partial y \partial x} & \dfrac{\partial^2 f}{\partial y^2} \end{vmatrix}$$

   Can you say if the graph has a max, min, or saddle at the critical point(s)?

   c. Find the slope of the tangent line to the graph of $f(x, y)$, in the direction of $\langle 1, 2 \rangle$, at the point $(0, 1)$.

   d. Find the volume under the graph of $f(x, y)$, and above the rectangle in the xy-plane with vertices at $(0, 0)$, $(1, 0)$, $(0, 2)$, and $(1, 2)$.

2. *Parameterize the portion of the graph of $z = 4 - x^2 - y^2$ that lies above the xy-plane.*

3. *Parameterize the curve that lies on a sphere of radius 1 such that $\theta = \phi$.*

4. *Let $\mathbf{W} = \langle xz^2, 0, xz^2 \rangle$. Calculate $\nabla \cdot \mathbf{W}$ and $\nabla \times \mathbf{W}$.*

5. *Let V be the volume in the first octant, inside the cylinder of radius 1, and below the plane $z = 1$. Integrate the function*

$$f(x, y) = 2\sqrt{1 + x^2 + y^2}$$

over V.

6. *Let C be the curve parameterized by the following:*

$$\phi(t) = (2 \cos t, 2 \sin t, t^2), \quad 0 \le t \le 2$$

Integrate the vector field $\langle 0, 0, x^2 + y^2 \rangle$ over C.

7. *Let $\mathbf{F}$ be the vector field $\langle 0, -z, 0 \rangle$. Let P be the portion of a paraboloid parameterized by*

$$\phi(r, \theta) = (r \cos \theta, r \sin \theta, r^2)$$

$$0 \le r \le 1, \quad 0 \le \theta \le \frac{\pi}{2}$$

Calculate $\int_P \mathbf{F} \cdot d\mathbf{S}$.

8. *Let $\mathbf{W} = \langle yz, xz, xy \rangle$. Let C be the curve parameterized by*

$$\phi(t) = \left(\frac{4t}{\pi} \cos t, \frac{4t}{\pi} \sin t, \frac{4t}{\pi} \right), \quad 0 \le t \le \frac{\pi}{4}$$

with the induced orientation. Evaluate $\int_C \mathbf{W} \cdot d\mathbf{s}$ without integrating!

9. *Let S be the can-shaped surface in $\mathbb{R}^3$ whose side is the cylinder of radius 1 (centered on the z-axis), and whose top and bottom are in the planes $z = 1$ and $z = 0$. Let $\mathbf{W} = \langle 0, 0, z^2 \rangle$. Use Gauss' Theorem to calculate $\int_S \mathbf{W} \cdot d\mathbf{S}$.*

Answers to Problems

Chapter 1: Functions of Multiple Variables

Problem 1

1. $f(3, 2) = 3^2 + 2^3 = 9 + 8 = 17$
2. $g(0, \frac{\pi}{2}) = \sin(0) + \cos(\frac{\pi}{2}) = 0 + 0 = 0$
3. $h(2, \frac{\pi}{2}) = 2^2 \sin(\frac{\pi}{2}) = (4)(1) = 4$

Problem 2

(b), (d)

Problem 3

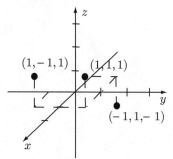

Problem 4

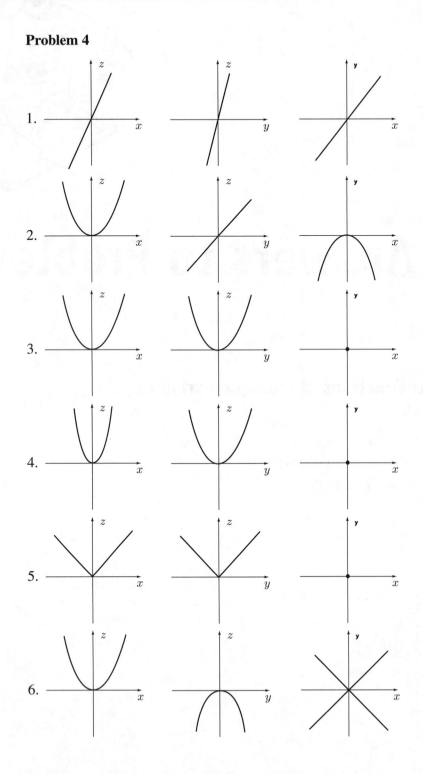

Problem 5

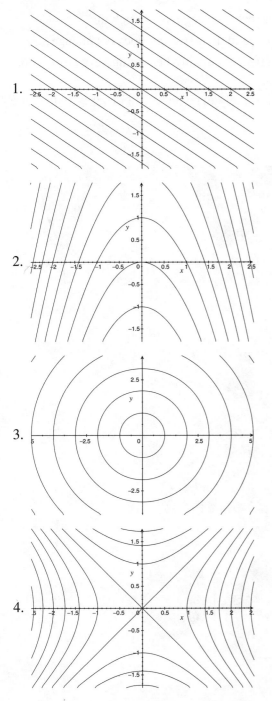

Problem 6

1. The circles get larger as z increases, but they begin to "bunch up."
2. The circles get larger as z gets larger, and their spacing does not change.
3. The circles get smaller as z gets larger, eventually becoming very close to the origin.
4. At $z = 0$ the level set consists of infinitely many circles. For $|z| > 1$ there are no level curves.

Problem 7

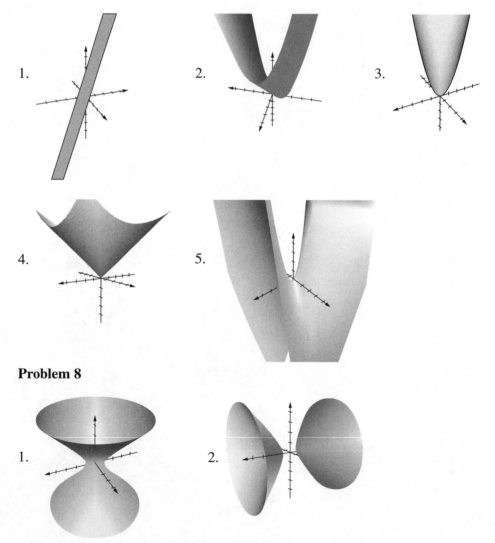

Problem 8

Problem 9

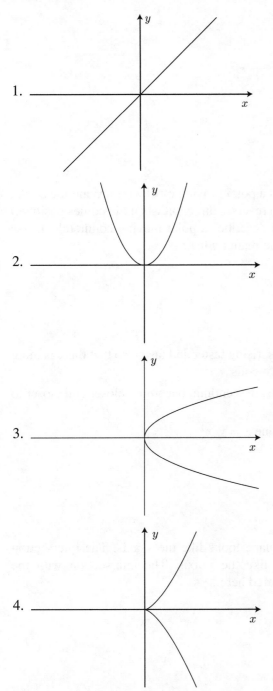

5.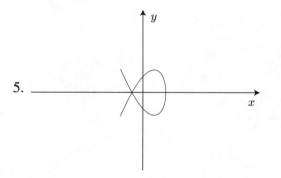

Problem 10

The function $(\cos t^2, \sin t^2)$ describes a point moving clockwise around the circle, and slowing down until $t = 0$. (It then reverses direction when t becomes positive.) In contrast, the function $(\cos t, \sin t)$ describes a point moving counterclockwise around the circle, at uniform speed, no matter what t is.

Problem 11

$c(t) = (t, f(t))$

Problem 12

1. The curve spirals around the z-axis, rising faster and faster, so that there is more and more space between successive coils.
2. The curve spirals around the z-axis, descending, becoming closer and closer to the xy-plane.
3. The curve spirals up around the cone $z = \sqrt{x^2 + y^2}$.

Chapter 1 Quiz

Problem 13

1. Left handed
2. (a) The intersection with the xy-plane looks like the x-axis. The intersection with the xz-plane also looks like the x-axis. The intersection with the yz-plane is the line $z = y$, pictured here.

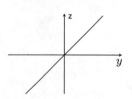

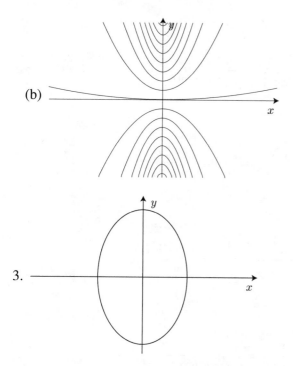

(b)

(c)

3.

Chapter 2: Fundamentals of Advanced Calculus

Problem 14

1. Along the y-axis we have

$$\lim_{(x,y)\to(0,0)} \frac{x^2}{x^2 + y^3} = \lim_{(x,y)\to(0,0)} \frac{0}{y^3} = 0$$

Along the x-axis

$$\lim_{(x,y)\to(0,0)} \frac{x^2}{x^2 + y^3} = \lim_{(x,y)\to(0,0)} \frac{x^2}{x^2} = 1$$

2. Along the y-axis

$$\lim_{(x,y)\to(0,0)} \frac{x^2 y}{x^3 + y^3} = \lim_{(x,y)\to(0,0)} \frac{0}{y^3} = 0$$

Along the line $y = x$

$$\lim_{(x,y)\to(0,0)} \frac{x^2 y}{x^3 + y^3} = \lim_{(x,y)\to(0,0)} \frac{x^3}{2x^3} = \frac{1}{2}$$

3. Along the y-axis

$$\lim_{(x,y)\to(0,0)} \frac{x+y}{\sqrt{x^2+y^2}} = \lim_{(x,y)\to(0,0)} \frac{y}{y} = 1$$

Along the line $y = x$

$$\lim_{(x,y)\to(0,0)} \frac{x+y}{\sqrt{x^2+y^2}} = \lim_{(x,y)\to(0,0)} \frac{2x}{\sqrt{2x}} = \sqrt{2}$$

4. Along the y-axis

$$\lim_{(x,y)\to(0,0)} \frac{x^2 y^2}{x^3+y^3} = \lim_{(x,y)\to(0,0)} \frac{0}{y^3} = 0$$

Along the curve $y = x^2$

$$\lim_{(x,y)\to(0,0)} \frac{x^2 y^2}{x^3+y^3} = \lim_{(x,y)\to(0,0)} \frac{x^6}{x^3+x^6} = 1$$

Problem 15

1. $y \neq x$ 2. $y \geq x^2$ 3. The domain is empty.

Problem 16

Yes. For small values of t the function $\sin(t)$ is approximately equal to t. The value of $x^2 + y^2$ is the square of the distance from (x, y) to $(0, 0)$, which is small for points near the origin.

Chapter 2 Quiz

Problem 17

1. On the x-axis we know $y = 0$, and so the function $f(x, y) = 0$ (as long as x is not also 0). Similarly, on the y-axis we know $x = 0$, and so again the function $f(x, y) = 0$. But when $x = y$ we have

$$f(x, y) = \frac{x \sin x}{2x^2}$$

To evaluate the limit of this as $x \to 0$ we use L'Hopital's rule twice

$$\lim_{x \to 0} \frac{x \sin x}{2x^2} = \lim_{x \to 0} \frac{\sin x + x \cos x}{4x}$$

$$= \lim_{x \to 0} \frac{2 \cos x - x \sin x}{4}$$

$$= \frac{1}{2}$$

Hence, the limit does not exist.

2. As long as both x and y are not zero, the function $\frac{x+y}{x+y} = 1$. Hence, the function $f(x, y) = 1$ for *every* value of x and y. The function $f(x, y)$ is thus a constant function, which is continuous everywhere.

3. The first problem we may encounter is a zero in the denominator, in which case $x - y^2 = 0$. The second problem is taking the log of a nonpositive number, in which case $\frac{1}{x-y^2} \leq 0$. But this second situation implies $x - y^2 \leq 0$, and hence includes the first.

   If $x - y^2 \leq 0$, then $x \leq y^2$. Hence, the domain of $f(x, y)$ is all the points (x, y) such that $x > y^2$.

Chapter 3: Derivatives

Problem 18

1. $\frac{\partial f}{\partial x}(2, 3) = 4$, $\frac{\partial f}{\partial y}(2, 3) = 2$

2. $\frac{\partial f}{\partial x}(2, 3) = \ln 3$, $\frac{\partial f}{\partial y}(2, 3) = \frac{2}{3}$

3. $\frac{\partial f}{\partial x}(2, 3) = \sqrt{6} + \frac{\sqrt{6}}{2}$, $\frac{\partial f}{\partial y}(2, 3) = \frac{\sqrt{6}}{3}$

Problem 19

1. $\frac{\partial f}{\partial x} = 2xy^3$, $\frac{\partial f}{\partial y} = 3x^2y^2$ 2. $\frac{\partial f}{\partial x} = \frac{1}{y}$, $\frac{\partial f}{\partial y} = \frac{-x}{y^2}$

Problem 20

$\frac{\partial f}{\partial x} = -1 + y^2$ and $\frac{\partial f}{\partial y} = 2xy - 2y$. Setting these both equal to zero gives the system of equations

$$0 = -1 + y^2$$

$$0 = 2xy - 2y$$

The first equation tells us $y^2 = 1$, or $y = \pm 1$. If $y = 1$ then the second equation becomes $0 = 2x - 2$, so $x = 1$. If $y = -1$ then the second equation becomes $0 = -2x + 2$, so $x = 1$. Hence, the solutions are $(1, 1)$ and $(1, -1)$.

Problem 21

First, note that $\phi(1) = (2, 1)$. So

$$\frac{\partial f}{\partial x}(\phi(1)) = \frac{\partial f}{\partial x}(2, 1) = 6, \text{ and}$$

$$\frac{\partial f}{\partial y}(\phi(1)) = \frac{\partial f}{\partial y}(2, 1) = -1$$

Now, note that $x(t) = 2t$, so $\frac{dx}{dt} = 2$. Similarly, $y(t) = t^2$, so $\frac{dy}{dt} = 2t$ and $\frac{dy}{dt}(1) = 2$. Finally, we compute

$$\frac{d}{dt} f(\phi(1)) = \frac{\partial f}{\partial x}(\phi(1))\frac{dx}{dt}(1) + \frac{\partial f}{\partial y}(\phi(1))\frac{dy}{dt}(1)$$

$$= 6 \cdot 2 + (-1) \cdot 2$$

$$= 10$$

Problem 22

First, note that $\frac{\partial f}{\partial x} = 2xy$ and $\frac{\partial f}{\partial y} = x^2$. Now we compute:

$$\frac{d}{dt} f(\phi(2)) = \frac{\partial f}{\partial x}(\phi(2))\frac{dx}{dt}(2) + \frac{\partial f}{\partial y}(\phi(2))\frac{dy}{dt}(2)$$

$$= \frac{\partial f}{\partial x}(1, 3)\frac{dx}{dt}(2) + \frac{\partial f}{\partial y}(1, 3)\frac{dy}{dt}(2)$$

$$= 2(1)(3)\frac{dx}{dt}(2) + 1^2\frac{dy}{dt}(2)$$

$$= (6)(-2) + (1)(1)$$

$$= -11$$

Problem 23

1. First, note that $x(1, 2) = 2$ and $y(1, 2) = 5$. Hence, if $(u, v) = (1, 2)$ then $(x, y) = (2, 5)$. Next, note that $\frac{\partial x}{\partial u} = v$ and $\frac{\partial y}{\partial u} = 1$, and so $\frac{\partial x}{\partial u}(1, 2) = 2$ and $\frac{\partial y}{\partial u}(1, 2) = 1$. We now compute:

$$\frac{\partial f}{\partial u}(1, 2) = \frac{\partial f}{\partial x}(2, 5)\frac{\partial x}{\partial u}(1, 2) + \frac{\partial f}{\partial y}(2, 5)\frac{\partial y}{\partial u}(1, 2)$$
$$= (2)(2) + (3)(1)$$
$$= 7$$

2. $x(2, 1) = 2$ and $y(2, 1) = 3$. Hence, if $(u, v) = (2, 1)$ then $(x, y) = (2, 3)$. Next, note that $\frac{\partial x}{\partial v} = u$ and $\frac{\partial y}{\partial v} = 2v$, and so $\frac{\partial x}{\partial v}(2, 1) = 2$ and $\frac{\partial y}{\partial v}(2, 1) = 2$. We now compute:

$$\frac{\partial f}{\partial v}(2, 1) = \frac{\partial f}{\partial x}(2, 3)\frac{\partial x}{\partial v}(2, 1) + \frac{\partial f}{\partial y}(2, 3)\frac{\partial y}{\partial v}(2, 1)$$
$$= (1)(2) + (-1)(2)$$
$$= 0$$

Problem 24

1. First, note that $x(\frac{\pi}{2}, \pi) = \frac{\pi}{2} + \pi = \frac{3\pi}{2}$ and $y(\frac{\pi}{2}, \pi) = \frac{\pi}{2} - \pi = -\frac{\pi}{2}$. Now,

$$f(x, y) = f\left(\frac{3\pi}{2}, -\frac{\pi}{2}\right) = \sin\left(\frac{3\pi}{2} - \frac{\pi}{2}\right) = \sin(\pi) = 0$$

2. First, note that $\frac{\partial f}{\partial x}(x, y) = \cos(x + y)$, so $\frac{\partial f}{\partial x}(\frac{3\pi}{2}, -\frac{\pi}{2}) = \cos(\pi) = -1$. Similarly, $\frac{\partial f}{\partial y}(x, y) = \cos(x + y)$, so $\frac{\partial f}{\partial y}(\frac{3\pi}{2}, -\frac{\pi}{2}) = \cos(\pi) = -1$. Finally, note that $\frac{\partial x}{\partial u} = 1$ and $\frac{\partial y}{\partial u} = 1$. Hence,

$$\frac{\partial f}{\partial u} = \frac{\partial f}{\partial x}\frac{\partial x}{\partial u} + \frac{\partial f}{\partial y}\frac{\partial y}{\partial u} = (-1)(1) + (-1)(1) = -2$$

3. Note that $\frac{\partial x}{\partial v} = 1$ and $\frac{\partial y}{\partial v} = -1$, so

$$\frac{\partial f}{\partial v} = \frac{\partial f}{\partial x}\frac{\partial x}{\partial v} + \frac{\partial f}{\partial y}\frac{\partial y}{\partial v} = (-1)(1) + (-1)(-1) = 0$$

Problem 25

1. $\dfrac{\partial^2 f}{\partial x^2} = 0, \quad \dfrac{\partial^2 f}{\partial y \, \partial x} = 1$

   $\dfrac{\partial^2 f}{\partial x \, \partial y} = 1, \quad \dfrac{\partial^2 f}{\partial y^2} = 0$

2. $\dfrac{\partial^2 f}{\partial x^2} = 2, \quad \dfrac{\partial^2 f}{\partial y \, \partial x} = 0$

   $\dfrac{\partial^2 f}{\partial x \, \partial y} = 0, \quad \dfrac{\partial^2 f}{\partial y^2} = -2$

3. $\dfrac{\partial^2 f}{\partial x^2} = -y^4 \sin(xy^2), \dfrac{\partial^2 f}{\partial y \, \partial x} = 2y \cos(xy^2) - 2xy^3 \sin(xy^2)$

   $\dfrac{\partial^2 f}{\partial x \, \partial y} = 2y \cos(xy^2) - 2xy^3 \sin(xy^2), \dfrac{\partial^2 f}{\partial y^2} = 2x \cos(xy^2) - 4x^2 y^2 \sin(xy^2)$

Chapter 3 Quiz

Problem 26

1. $\frac{\partial f}{\partial x} = 2xy + 3x^2 y^2$ and $\frac{\partial f}{\partial y} = x^2 + 2x^3 y$.

2. $\begin{aligned} f(\phi(t)) &= (t^2)^2 (t-1) + (t^2)^3 (t-1)^2 \\ &= t^4 (t-1) + t^6 (t^2 - 2t + 1) \\ &= t^8 - 2t^7 + t^6 + t^5 - t^4 \end{aligned}$

3. First note from Question 1 that $\frac{\partial f}{\partial x}(1, 1) = 5$ and $\frac{\partial f}{\partial y}(1, 1) = 3$. Now,

$$\begin{aligned} \frac{df(\psi(t))}{dt} &= \frac{\partial f}{\partial x}\frac{dx}{dt} + \frac{\partial f}{\partial y}\frac{dy}{dt} \\ &= (5)(3) + (3)(1) \\ &= 18 \end{aligned}$$

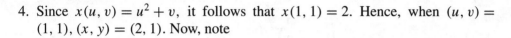

4. Since $x(u, v) = u^2 + v$, it follows that $x(1, 1) = 2$. Hence, when $(u, v) = (1, 1)$, $(x, y) = (2, 1)$. Now, note

$$\frac{\partial f}{\partial x}(2, 1) = 4 + 12 = 16$$

$$\frac{\partial f}{\partial y}(2, 1) = 4 + 16 = 20, \text{ and}$$

$$\frac{\partial x}{\partial u}(1, 1) = 2(1) = 2$$

Now, we have

$$12 = \frac{\partial f}{\partial u}$$

$$= \frac{\partial f}{\partial x}\frac{\partial x}{\partial u} + \frac{\partial f}{\partial y}\frac{\partial y}{\partial u}$$

$$= (16)(2) + 20\frac{\partial y}{\partial u}$$

And so,

$$12 = 32 + 20\frac{\partial y}{\partial u}$$

from which it follows that

$$\frac{\partial y}{\partial u} = -1$$

Chapter 4: Integration

Problem 27

1. $$\int_0^1 \int_2^3 x + xy^2 \, dx \, dy = \int_0^1 \frac{1}{2}x^2 + \frac{1}{2}x^2 y^2 \Big|_{x=2}^3 \, dy$$

$$= \int_0^1 \frac{9}{2} + \frac{9}{2}y^2 - 2 - 2y^2 \, dy$$

$$= \int_0^1 \frac{5}{2} + \frac{5}{2}y^2 \, dy$$

$$= \frac{5}{2}y + \frac{5}{6}y^3 \Big|_0^1$$

$$= \frac{5}{2} + \frac{5}{6}$$

$$= \frac{10}{3}$$

2.
$$\int_{-1}^1 \int_0^1 x^2 y^2 \, dy \, dx = \int_{-1}^1 \frac{1}{3}x^2 y^3 \Big|_{y=0}^1 \, dx$$

$$= \int_{-1}^1 \frac{1}{3}x^2 \, dx$$

$$= \frac{1}{9}x^3 \Big|_{-1}^1$$

$$= \frac{1}{9} - \left(-\frac{1}{9}\right)$$

$$= \frac{2}{9}$$

3.
$$\int_0^{\frac{\pi}{2}} \int_0^{\frac{\pi}{2}} \cos(x+y) \, dx \, dy = \int_0^{\frac{\pi}{2}} \sin(x+y) \Big|_{x=0}^{\frac{\pi}{2}} \, dy$$

$$= \int_0^{\frac{\pi}{2}} \sin\left(\frac{\pi}{2} + y\right) - \sin y \, dy$$

$$= -\cos\left(\frac{\pi}{2} + y\right) - \cos y \Big|_0^{\frac{\pi}{2}}$$

$$= -\cos \pi - \cos \frac{\pi}{2} + \cos \frac{\pi}{2} + \cos 0$$

$$= 2$$

Problem 28

$$\text{Volume} = \int_{-1}^{0} \int_{-1}^{2} x^3 y \, dx \, dy$$

$$= \int_{-1}^{0} \frac{1}{4} x^4 y \bigg|_{x=-1}^{2} \, dy$$

$$= \int_{-1}^{0} 4y - \frac{1}{4} y \, dy$$

$$= 2y^2 - \frac{1}{8} y^2 \bigg|_{-1}^{0}$$

$$= -2 + \frac{1}{8}$$

$$= -\frac{15}{8}$$

Problem 29

$$\text{Volume} = \int_{0}^{1} \int_{0}^{1} x^n y^m \, dx \, dy$$

$$= \int_{0}^{1} \frac{1}{n+1} x^{n+1} y^m \bigg|_{x=0}^{1} \, dy$$

$$= \int_{0}^{1} \frac{1}{n+1} y^m \, dy$$

$$= \frac{1}{(n+1)(m+1)} y^{m+1} \bigg|_{0}^{1}$$

$$= \frac{1}{(n+1)(m+1)}$$

Problem 30

To find the answer just set $y = 1$ and integrate with respect to x

$$\text{Area} = \int_0^2 e^{-x \cdot 1} dx = -e^{-x} \Big|_0^2 = 1 - \frac{1}{e^2}$$

Problem 31

1. $\displaystyle\int_0^1 \int_0^{y^2} 2xy^3 \, dx \, dy = \int_0^1 x^2 y^3 \Big|_{x=0}^{y^2} dy$

$$= \int_0^1 (y^2)^2 y^3 \, dy$$

$$= \int_0^1 y^7 \, dy$$

$$= \frac{1}{8} y^8 \Big|_0^1$$

$$= \frac{1}{8}$$

2. $\displaystyle\int_0^2 \int_x^{2x} e^{x+y} \, dy \, dx = \int_0^2 e^{x+y} \Big|_x^{2x} dx$

$$= \int_0^2 e^{3x} - e^{2x} \, dx$$

$$= \frac{1}{3} e^{3x} - \frac{1}{2} e^{2x} \Big|_0^2$$

$$= \frac{1}{3} e^6 - \frac{1}{2} e^4 + \frac{1}{6}$$

Problem 32

The graph of $y = x^2 - x - 2$ is depicted below.

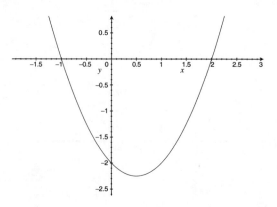

The x-intercepts of the graph are $x = -1$ and $x = 2$. Fixing a value of x in between these two numbers, y will range from $x^2 - x - 2$ to 0. This tells us how to set up the integral

$$\int_{-1}^{2} \int_{x^2-x-2}^{0} x^2 \, dy \, dx = \int_{-1}^{2} x^2 y \Big|_{x^2-x-2}^{0} \, dx$$

$$= \int_{-1}^{2} -x^2(x^2 - x - 2) \, dx$$

$$= \int_{-1}^{2} -x^4 + x^3 + 2x^2 \, dx$$

$$= -\frac{1}{5}x^5 + \frac{1}{4}x^4 + \frac{2}{3}x^3 \Big|_{-1}^{2}$$

$$= -\frac{32}{5} + \frac{16}{4} + \frac{16}{3} - \frac{1}{5} - \frac{1}{4} + \frac{2}{3}$$

$$= -\frac{33}{5} + \frac{15}{4} + \frac{18}{3}$$

Problem 33

The region R is depicted below

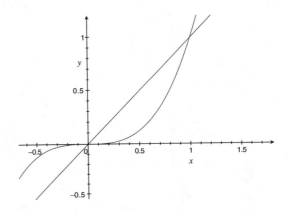

- The range of possible x-values is 0 to 1. If we fix a value of x between these numbers, then y can vary from x^3 to x. So the integral can be set up as

$$\int_0^1 \int_{x^3}^x f(x, y)\, dy\, dx$$

- The range of possible y-values is also 0 to 1. If we fix a value of y between these numbers, then x can vary from y to $\sqrt[3]{y}$. Hence, the integral can be set up as

$$\int_0^1 \int_y^{\sqrt[3]{y}} f(x, y)\, dx\, dy$$

Problem 34

The original limits of integration tell us the shape of the region over which we are integrating the function $\sin(y^3)$. The outside limits tell us that the domain of integration lies between the vertical lines $x = 0$ and $x = \pi^2$. Fixing a value of x between these numbers, y can vary from $\sqrt{x}$ to π (the inside limits of integration).

This tells us the shape of the domain of integration is

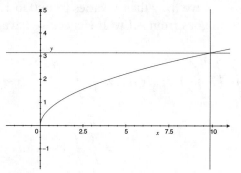

Now notice from the figure that the possible range of y values is from 0 to π. Fixing a value of y between these numbers, x can range from 0 to y^2. Hence, we may rewrite the integral and evaluate as follows:

$$\int_0^\pi \int_0^{y^2} \sin(y^3) \, dx \, dy = \int_0^\pi x \sin(y^3) \big|_{x=0}^{y^2} \, dy$$

$$= \int_0^\pi y^2 \sin(y^3) \, dy$$

$$= -\frac{1}{3} \cos(y^3) \bigg|_0^\pi$$

$$= -\frac{1}{3}(\cos \pi^3 - \cos 0)$$

$$= \frac{1}{3}(1 - \cos \pi^3)$$

Problem 35

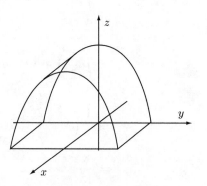

The solid is pictured above. To find its volume, note that if we fix x and y then z varies from 0 to $1 - y^2$. If we fix y then x varies from 0 to 1. Finally, the range of values that y can take on goes from -1 to 1. Hence, we have the triple integral:

$$\int_{-1}^{1} \int_{0}^{1} \int_{0}^{1-y^2} 1 \, dz \, dx \, dy = \int_{-1}^{1} \int_{0}^{1} z \Big|_{0}^{1-y^2} dx \, dy$$

$$= \int_{-1}^{1} \int_{0}^{1} 1 - y^2 \, dx \, dy$$

$$= \int_{-1}^{1} 1 - y^2 \, dy$$

$$= y - \frac{1}{3} y^3 \Big|_{-1}^{1}$$

$$= \frac{4}{3}$$

Problem 36

The volume can be computed with a triple integral as follows:

$$\int_{-1}^{1} \int_{0}^{1-y^2} \int_{0}^{1-x-y^2} 1 \, dz \, dx \, dy = \int_{-1}^{1} \int_{0}^{1-y^2} 1 - x - y^2 \, dx \, dy$$

$$= \int_{-1}^{1} x - \frac{1}{2} x^2 - xy^2 \Big|_{0}^{1-y^2} dy$$

$$= \int_{-1}^{1} \frac{1}{2} (1 - y^2)^2 \, dy$$

$$= \frac{1}{2} \left(y - \frac{2}{3} y^3 + \frac{1}{5} y^5 \right) \Big|_{-1}^{1}$$

$$= \frac{8}{15}$$

Problem 37

$$\int_{-1}^{1} \int_{-\sqrt{1-y^2}}^{\sqrt{1-y^2}} \int_{0}^{1-x^2-y^2} 1 \, dz \, dx \, dy$$

Problem 38

$$\int_{-1}^{1} \int_{-\sqrt{1-z^2}}^{\sqrt{1-z^2}} \int_{-\sqrt{1-z^2-x^2}}^{\sqrt{1-z^2-x^2}} 1 \, dy \, dx \, dz$$

Chapter 4 Quiz

Problem 39

1.(a)
$$\int_{1}^{2} \int_{2}^{3} \cos(2x+y) \, dx \, dy = \int_{1}^{2} \frac{1}{2} \sin(2x+y) \Big|_{2}^{3} \, dy$$

$$= \frac{1}{2} \int_{1}^{2} \sin(6+y) - \sin(4+y) \, dy$$

$$= \frac{1}{2} \left(\cos(4+y) - \cos(6+y) \right) \big|_{1}^{2}$$

$$= \frac{1}{2} (\cos 6 - \cos 7)$$

(b) The trick with this integral is to reverse the order of integration:

$$\int_{0}^{1} \int_{x}^{1} \sqrt{1+y^2} \, dy \, dx = \int_{0}^{1} \int_{0}^{y} \sqrt{1+y^2} \, dx \, dy$$

$$= \int_{0}^{1} y\sqrt{1+y^2} \, dy$$

$$= \int_{1}^{2} \frac{1}{2} \sqrt{u} \, du$$

$$= \frac{1}{3}u^{\frac{3}{2}}\Big|_1^2$$

$$= \frac{1}{3}(2^{\frac{3}{2}} - 1)$$

2. The domain of integration is a disk of radius 1 in the xy-plane. Hence, if we fix x then y can vary from $-\sqrt{1-x^2}$ to $\sqrt{1-x^2}$. The biggest and smallest possible values of x are -1 and 1. Hence, the desired volume can be computed by the integral

$$\int_{-1}^{1} \int_{-\sqrt{1-x^2}}^{\sqrt{1-x^2}} \sqrt{1 - x^2 - y^2} \, dy \, dx$$

Chapter 5: Cylindrical and Spherical Coordinates

Problem 40

1. $(\sqrt{2}, \sqrt{2}, -1)$ 2. $(0, 0, 3)$ 3. $(2, 2\sqrt{3}, 0)$

Problem 41

1. The xz-plane.   2. The yz-plane.   3. The xy-plane.
4. A right circular cone centered on the z-axis.   5. A sphere of radius 1.

Problem 42

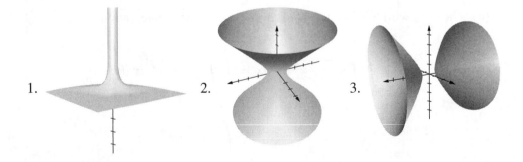

1.   2.   3.

Problem 43

1. $(0, 0, 0)$ 2. $(-3, 0, 0)$ 3. $(\sqrt{6}, \sqrt{6}, 2)$

Problem 44

1. A single point at the origin.
2. The xy-plane.
3. The negative z-axis.
4. A cylinder of radius 2, centered on the z-axis. (Recall that $\rho \sin \phi = r$, so this is just the graph of the cylindrical equation $r = 2$.)
5. A horizontal plane at height 2. (Recall that $\rho \cos \phi = z$, so this is just the graph of the rectangular equation $z = 2$.)

Problem 45

1. 2.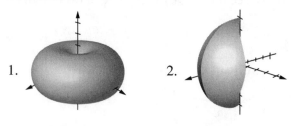

Chapter 5 Quiz

Problem 46

1.(a) $x = r \cos \theta = \cos \dfrac{\pi}{6} = \dfrac{\sqrt{3}}{2}$

$y = r \sin \theta = \sin \dfrac{\pi}{6} = \dfrac{1}{2}$

$z = z = 2$

(b) $x = \rho \sin \phi \cos \theta = 2 \sin \dfrac{\pi}{4} \cos \dfrac{\pi}{6} = 2 \dfrac{\sqrt{2}}{2} \dfrac{\sqrt{3}}{2} = \dfrac{\sqrt{6}}{2}$

$y = \rho \sin \phi \sin \theta = 2 \sin \dfrac{\pi}{4} \sin \dfrac{\pi}{6} = 2 \dfrac{\sqrt{2}}{2} \dfrac{1}{2} = \dfrac{\sqrt{2}}{2}$

$z = \rho \cos \phi = 2 \cos \dfrac{\pi}{4} = 2 \dfrac{\sqrt{2}}{2} = \sqrt{2}$

2.(a) In polar coordinates the equation $r = \cos \theta$ describes a circle of radius $\frac{1}{2}$, centered on the point $(\frac{1}{2}, 0)$. In cylindrical coordinates, we see the same picture no matter what z is. Hence, the graph is the cylinder pictured below.

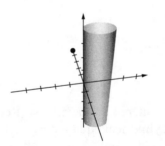

(b) Consider the sphere of radius 1. On this sphere the points where $\theta = \phi$ forms a curve which starts at the top (where $z = 1$), ends at the bottom (where $z = -1$), and passes through the point where $y = 1$. Now we see the same picture no matter what ϕ is, so we see this curve on a sphere of every radius, as in the figure below.

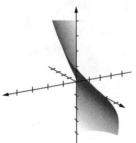

Chapter 6: Parameterizations

Problem 47

1. $\psi(x, y) = (x, y, x^2)$
2. $\psi(\theta, z) = (\theta^2 \cos \theta, \theta^2 \sin \theta, z)$
3. $\psi(\theta, \phi) = (\theta^2 \sin \phi \cos \theta, \theta^2 \sin \phi \sin \theta, \theta^2 \cos \phi)$

Problem 48

1. $\phi = \theta$ 2. $y = x + xz$ 3. $z = \sin r$

Problem 49

1. $0 \le \theta \le \frac{\pi}{2}, \quad 0 \le \phi \le \pi$ 2. $\frac{\pi}{2} \le \theta\pi, \quad \frac{\pi}{2} \le \phi\pi$

Problem 50

1. A portion of a cone that lies above a square.
2. A portion of a cone that lies above a circle.
3. A portion of a cone that lies above half of a circle.

Problem 51

1. Start with the graph of $x = y^2 + z$. Now stretch this graph in the y-direction by a factor of 2. Finally, translate the entire picture -1 unit in the z-direction.
2. Start with the graph of $z = x + \sin y$. Now reflect in the plane $y = x$, switching the roles of x and y.
3. Start with the cylinder $r = 1$. Stretch in the x-direction by 2 and the y-direction by 3, so that cross sections are now ellipses. Now move the central axis 1 unit in the y-direction and -1 unit in the z-direction. Note that this last translation does nothing to the shape.

Problem 52

If the curves shown were circles, with the one corresponding to $z = 1$ having radius 1, then the uniform spacing of the level curves would tell us that the shape is a cone. Such a cone would be the graph of the cylindrical equation $z = r$, and would thus be parameterized by

$$\Psi(r, \theta) = (r \cos \theta, r \sin \theta, r)$$

$$0 \le r \le 1, \quad 0 \le \theta \le 2\pi$$

The actual level curves we are given are ellipses. These can be obtained from circular level curves by multiplying the x- and y-coordinates by a factor which will stretch them the appropriate amount:

$$\Psi(r, \theta) = (2r \cos \theta, 3r \sin \theta, r)$$

$$0 \le r \le 1, \quad 0 \le \theta \le 2\pi$$

Problem 53

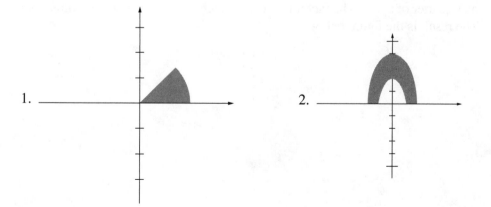

1.

2.

Problem 54

1. The region between spheres of radii 1 and 2, and above the xy-plane.
2. The region inside a cylinder of height 2 and radius 1, between the planes $z = 1$ and $z = 3$.

Problem 55

$$\Psi(r, \theta, t) = (r \cos \theta, r \sin \theta, tr)$$

$$0 \le r \le 1, \quad 0 \le \theta \le 2\pi, \quad 0 \le t \le 1$$

Chapter 6 Quiz

Problem 56

1. For each point on a cylinder of radius 2, we know $r = 2$. If such points are also on the graph of $z = 2r$, then $z = 2(2) = 4$. Hence, the problem is asking for the points of a cylinder of radius 2 for which $0 \le z \le 4$. This is most easily done by utilizing cylindrical coordinates:

$$\Psi(\theta, z) = (2 \cos \theta, 2 \sin \theta, z)$$

$$0 \le \theta \le 2\pi, \quad 0 \le z \le 2$$

2. If the 2 were not at the x-coordinate then this parameterization would look just like spherical coordinates. The restrictions on ρ, θ, and ϕ would then specify one-quarter of a ball. The factor of 2 then stretches this shape in the x-direction. The result is the figure below.

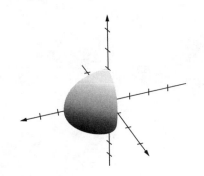

Chapter 7: Vectors and Gradients

Problem 57

1. $V + W = \langle 7, 7 \rangle, -V = \langle -1, -6 \rangle, V - W = \langle -5, 5 \rangle$
2. $V + W = \langle 1, 2 \rangle, -V = \langle 0, 0 \rangle, V - W = \langle -1, -2 \rangle$
3. $V + W = \langle 0, -1 \rangle, -V = \langle 1, -1 \rangle, V - W = \langle -2, 3 \rangle$

Problem 58

1. $\sqrt{13}$ 2. $\sqrt{10}$

Problem 59

The magnitude of $\langle 2, 1 \rangle$ is $\sqrt{2^2 + 1^2} = \sqrt{5}$. To get a unit vector that points in the same direction we just divide by the magnitude:

$$\frac{1}{\sqrt{5}} \langle 2, 1 \rangle = \left\langle \frac{2\sqrt{5}}{5}, \frac{\sqrt{5}}{5} \right\rangle$$

Problem 60

The vector $\langle 5, 12 \rangle$ lies on a line of slope $\frac{12}{5}$. A perpendicular line will have a slope which is the negative reciprocal of this, $\frac{-5}{12}$. A vector that lies in this line is $\langle -12, 5 \rangle$. To get a unit vector that points in the same direction, we divide by the magnitude:

$$\frac{1}{\sqrt{12^2 + 5^3}} \langle -12, 5 \rangle = \frac{1}{13} \langle -12, 5 \rangle = \left\langle -\frac{12}{13}, \frac{5}{13} \right\rangle$$

Problem 61

1. $(2)(-3) + (4)(1) = -2$
2. $(0)(5) + (7)(2) = 14$
3. $(-2)(6) + (-1)(-3) = -9$

Problem 62

1. $\cos\theta = \dfrac{\langle 2, 4 \rangle \cdot \langle -3, 1 \rangle}{|\langle 2, 4 \rangle||\langle -3, 1 \rangle|} = \dfrac{-2}{\sqrt{20}\sqrt{10}} = -\dfrac{\sqrt{2}}{10}$

2. $\cos\theta = \dfrac{\langle -2, -1 \rangle \cdot \langle 6, -3 \rangle}{|\langle -2, -1 \rangle||\langle 6, -3 \rangle|} = \dfrac{-9}{\sqrt{5}\sqrt{45}} = -\dfrac{6}{15}$

Problem 63

$V \cdot W = |V||W| \cos \theta = 9 \cos \theta$. So $V \cdot W$ is largest when the value of $\cos \theta$ is largest. Since the largest possible value of $\cos \theta$ is one (when $\theta = 0$), the largest possible value of $V \cdot W$ is 9. Similarly, the smallest possible value for $\cos \theta$ is -1, so the smallest possible value for $V \cdot W$ is -9. If V and W are perpendicular then $\cos \theta = 0$, so the smallest possibility for $|V \cdot W|$ is zero.

Problem 64

1. No, $V \cdot W = 4 \neq 0$.   2. Yes, $V \cdot W = 0$.
3. Yes, $V \cdot W = 0$.   4. No, $V \cdot W = -9 \neq 0$.

Problem 65

1. • $\nabla f(x, y) = \langle \ln y, \frac{x}{y} \rangle$
   • $\nabla f(1, 1) \cdot \langle \frac{3}{5}, \frac{4}{5} \rangle = \langle 0, 1 \rangle \cdot \langle \frac{3}{5}, \frac{4}{5} \rangle = \frac{4}{5}$
   • The direction of the maximum rate of change at $(1, 1)$ is $\nabla f(1, 1) = \langle 0, 1 \rangle$. This is already a unit vector.
   • $|\nabla f(1, 1)| = 1$
   • Such a line is perpendicular to $\nabla f(1, 1) = \langle 0, 1 \rangle$. Such a vector is $\langle 1, 0 \rangle$.

2. • $\nabla f(x, y) = \langle 2, 3 \rangle$
   • $\nabla f(1, 1) \cdot \langle \frac{3}{5}, \frac{4}{5} \rangle = \langle 2, 3 \rangle \cdot \langle \frac{3}{5}, \frac{4}{5} \rangle = \frac{18}{5}$
   • The direction of maximum rate of change is $\nabla f(1, 1) = \langle 2, 3 \rangle$. A unit vector that points in this direction is found by dividing by $\sqrt{13}$, the magnitude of this vector. The result is $\langle \frac{2\sqrt{13}}{13}, \frac{3\sqrt{13}}{13} \rangle$.
   • $|\nabla f(1, 1)| = \sqrt{13}$
   • A line perpendicular to the one containing $\nabla f(1, 1) = \langle 2, 3 \rangle$ would have slope $\frac{-2}{3}$. A vector in this line is $\langle -3, 2 \rangle$. A unit vector pointing in the same direction is $\langle -\frac{3\sqrt{13}}{13}, \frac{2\sqrt{13}}{13} \rangle$.

3. • $\nabla f(x, y) = \langle 2xy + y^3, x^2 + 3xy^2 \rangle$
   • $\nabla f(1, 1) \cdot \langle \frac{3}{5}, \frac{4}{5} \rangle = \langle 3, 4 \rangle \cdot \langle \frac{3}{5}, \frac{4}{5} \rangle = 5$
   • The direction of maximum rate of change is $\nabla f(1, 1) = \langle 3, 4 \rangle$. A unit vector that points in the same direction is $\langle \frac{3}{5}, \frac{4}{5} \rangle$.
   • $|\nabla f(1, 1)| = 5$
   • A line perpendicular to the one containing $\nabla f(1, 1) = \langle 3, 4 \rangle$ would have slope $\frac{-3}{4}$. A vector in this line is $\langle -4, 3 \rangle$. A unit vector pointing in the same direction is $\langle -\frac{4}{5}, \frac{3}{5} \rangle$.

Problem 66

1. First, we find the critical points by looking for places where $\nabla f(x, y) = \langle 0, 0 \rangle$. Setting the vector $\nabla f(x, y) = \langle y + 2, x - 3 \rangle$ equal to $\langle 0, 0 \rangle$, we see that x must be 3 and y must be -2. So the only critical point is at $(3, -2)$. Now note that $D(x, y) = -1$ for all (x, y). In particular, $D(3, -2) = -1 < 0$, so the critical point corresponds to a saddle.

2. Setting $\nabla f(x, y) = \langle 3x^2 - y, -x + 2y \rangle$ equal to $\langle 0, 0 \rangle$, we see that the coordinates of any critical point must satisfy the system

$$3x^2 - y = 0$$
$$-x + 2y = 0$$

The second equation tells us $x = 2y$. Plugging this into the first equation gives $3(2y)^2 - y = 0$. Solving for y then gives solutions at $y = 0$ and $y = \frac{1}{12}$. Plugging these numbers back into $x = 2y$ gives us x values of 0 and $\frac{1}{6}$. So the coordinates of the critical points are $(0, 0)$ and $(\frac{1}{6}, \frac{1}{12})$.
We now compute

$$D(x, y) = (6x)(2) - (-1)^2 = 12x - 1$$

We conclude $D(0, 0) = -1 < 0$, so $(0, 0)$ is a saddle. On the other hand, $D(\frac{1}{6}, \frac{1}{12}) = 2 - 1 = 1 > 0$, so there is either as maximum or a minimum at $(\frac{1}{6}, \frac{1}{12})$. As $\frac{\partial^2 f}{\partial x^2}(\frac{1}{6}, \frac{1}{12}) = 1 > 0$, it must be a minimum.

Problem 67

All second partials are equal to $-\sin(x + y)$, so

$$D(x, y) = \frac{\partial^2 f}{\partial x^2} \frac{\partial^2 f}{\partial y^2} - \left(\frac{\partial^2 f}{\partial x \, \partial y}\right)^2 = \sin^2(x + y) - \sin^2(x + y) = 0$$

Inspection of the graph, pictured below, shows that there are infinitely many maxima and minima, but no saddles.

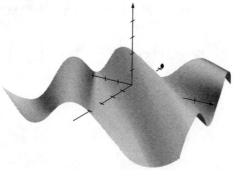

Problem 68

Recall that $D(x, y) = \frac{\partial^2 f}{\partial x^2} \frac{\partial^2 f}{\partial y^2} - (\frac{\partial^2 f}{\partial x \partial y})^2$. Since $(\frac{\partial^2 f}{\partial x \partial y})^2$ is always bigger than or equal to zero, $\frac{\partial^2 f}{\partial x^2} \frac{\partial^2 f}{\partial y^2} - (\frac{\partial^2 f}{\partial x \partial y})^2$ can only be positive if $\frac{\partial^2 f}{\partial x^2} \frac{\partial^2 f}{\partial y^2}$ is positive. But if the product of two numbers is positive, and you know the first of those two is also positive, then the second must also be positive.

Problem 69

1. First we look for critical points in the interior of D by setting the partials equal to zero:

$$\frac{\partial f}{\partial x} = 2x = 0$$

$$\frac{\partial f}{\partial y} = 2y + 2 = 0$$

From the first equation $x = 0$, and from the second $y = -1$. So $(0, -1)$ is a critical point. (To find out what type we look at the matrix of second partials:

$$\begin{vmatrix} 2 & 0 \\ 0 & 2 \end{vmatrix} = 4 > 0$$

Since $\frac{\partial^2 f}{\partial x^2} = 2$ is also greater than 0, the point $(0, -1)$ represents a minimum.)
We now look for minima on the boundary of D using the method of LaGrange multipliers. The boundary of D satisfies the equation $g(x, y) = x^2 + \frac{y^2}{4} = 1$. Hence, we must look for points (x_0, y_0) that satisfy this and

$$\nabla f(x_0, y_0) = \lambda \nabla g(x_0, y_0)$$

Note that $\nabla f = \langle 2x, 2y + 2 \rangle$ and $\nabla g = \langle 2x, \frac{y}{2} \rangle$. Hence, we get two equations:

$$2x = \lambda(2x), \quad 2y + 2 = \lambda \left(\frac{1}{2} y \right)$$

The first equation tells us either $\lambda = 1$ or $x = 0$. If $\lambda = 1$ then the second equation becomes

$$2y + 2 = \frac{y}{2}$$

Solving for y then gives $y = -\frac{4}{3}$. However, if we try to find corresponding values for x by plugging this into $x^2 + \frac{y^2}{4} = 1$ (the equation for the boundary of D) we get imaginary answers.

If $x = 0$, then the equation $x^2 + \frac{y^2}{4} = 1$ gives us $y = \pm 2$. So on the boundary of D the only potential minima are at $(0, 2)$ and $(0, -2)$.

2. To check which critical point represents the minimum of f we just plug them in:

$$f(0, -1) = 1 - 2 - 1 = -2$$
$$f(0, 2) = 4 + 4 - 1 = 7$$
$$f(0, -2) = 4 - 4 - 1 = -1$$

So the smallest value attained by $f(x, y)$ on D is -2, and this happens at $(0, -1)$.

Problem 70

1. $\begin{vmatrix} 1 & 3 \\ -1 & 2 \end{vmatrix} = 5$ 2. $\begin{vmatrix} 1 & 6 \\ 1 & 1 \end{vmatrix} = -5$ 3. $\begin{vmatrix} 2 & 3 \\ 6 & 9 \end{vmatrix} = 0$

Problem 71

Since V and W are parallel there is a number k such that $V = kW$. So if $V = \langle a, b \rangle$ then $W = \langle ka, kb \rangle$. We now compute

$$\begin{vmatrix} a & b \\ ka & kb \end{vmatrix} = akb - bka = 0$$

Problem 72

1. $\begin{vmatrix} 1 & 2 & 3 \\ 1 & 0 & 2 \\ -2 & 2 & -3 \end{vmatrix} = 1(0 - 4) - 2(-3 + 4) + 3(2 - 0) = -4 - 2 + 6 = 0$

2. $\begin{vmatrix} 0 & 1 & 3 \\ -1 & 2 & 1 \\ 2 & 0 & -1 \end{vmatrix} = 0(-2 - 0) - 1(1 - 2) + 3(0 - 4) = 0 + 1 - 12 = -11$

Problem 73

1. $\begin{vmatrix} \mathbf{i} & \mathbf{j} & \mathbf{k} \\ 1 & 2 & 3 \\ -1 & 0 & 1 \end{vmatrix} = |\langle 2, -4, 2 \rangle| = \sqrt{4 + 16 + 4} = 2\sqrt{6}$

2. $\begin{vmatrix} \mathbf{i} & \mathbf{j} & \mathbf{k} \\ 1 & 1 & 0 \\ 1 & 0 & 1 \end{vmatrix} = |\langle 1, -1, -1 \rangle| = \sqrt{1 + 1 + 1} = \sqrt{3}$

3. $\begin{vmatrix} \mathbf{i} & \mathbf{j} & \mathbf{k} \\ 1 & 2 & 3 \\ 3 & 1 & 2 \end{vmatrix} = |\langle 1, 7, -5 \rangle| = \sqrt{1 + 49 + 25} = 5\sqrt{3}$

Problem 74

1. First we compute the cross product:

$$\begin{vmatrix} \mathbf{i} & \mathbf{j} & \mathbf{k} \\ 1 & 2 & 0 \\ 1 & 1 & 1 \end{vmatrix} = \langle 2, -1, -1 \rangle$$

The magnitude of this vector is $\sqrt{4 + 1 + 1} = \sqrt{6}$. So a unit vector that points in the right direction is

$$\frac{\langle 2, -1, -1 \rangle}{\sqrt{6}} = \left\langle \frac{\sqrt{6}}{3}, -\frac{\sqrt{6}}{6}, -\frac{\sqrt{6}}{6} \right\rangle$$

2. The cross product is

$$\begin{vmatrix} \mathbf{i} & \mathbf{j} & \mathbf{k} \\ 1 & 1 & 2 \\ 1 & 1 & 1 \end{vmatrix} = \langle -1, 1, 0 \rangle$$

The magnitude of this is $\sqrt{2}$. So a unit vector that points in the right direction is

$$\frac{\langle -1, 1, 0 \rangle}{\sqrt{2}} = \left\langle \frac{\sqrt{2}}{2}, -\frac{\sqrt{2}}{2}, 0 \right\rangle$$

Problem 75

Let $V = \langle a, b, c \rangle$ and $W = \langle c, d, e \rangle$. We now compute $V \times W$ and $W \times V$:

$$V \times W = \begin{vmatrix} \mathbf{i} & \mathbf{j} & \mathbf{k} \\ a & b & c \\ d & e & f \end{vmatrix} = \mathbf{i}(bf - ce) - \mathbf{j}(af - cd) + \mathbf{k}(ae - bd)$$

$$W \times V = \begin{vmatrix} \mathbf{i} & \mathbf{j} & \mathbf{k} \\ d & e & f \\ a & b & c \end{vmatrix} = \mathbf{i}(ce - bf) - \mathbf{j}(cd - af) + \mathbf{k}(bd - ae)$$

Problem 76

Let $U = \langle a, b, c \rangle$, $V = \langle d, e, f \rangle$ and $W = \langle g, h, i \rangle$. Now compute:

$$U \cdot (V \times W) = \langle a, b, c \rangle \cdot \begin{vmatrix} \mathbf{i} & \mathbf{j} & \mathbf{k} \\ d & e & f \\ g & h & i \end{vmatrix}$$

$$= \langle a, b, c \rangle \cdot \left\langle \begin{vmatrix} e & f \\ h & i \end{vmatrix}, -\begin{vmatrix} d & f \\ g & i \end{vmatrix}, \begin{vmatrix} d & e \\ g & h \end{vmatrix} \right\rangle$$

$$= a\begin{vmatrix} e & f \\ h & i \end{vmatrix} - b\begin{vmatrix} d & f \\ g & i \end{vmatrix} + c\begin{vmatrix} d & e \\ g & h \end{vmatrix}$$

$$= \begin{vmatrix} a & b & c \\ d & e & f \\ g & h & i \end{vmatrix}$$

Problem 77

Inspection of the following figure reveals that $|W| \sin \theta$ is the height of the parallelogram, and $|V|$ is the length of the base. The area is just the product of the base and the height.

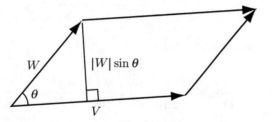

Chapter 7 Quiz

Problem 78

1.(a) The gradient of $f(x, y)$ at (x, y) is

$$\nabla f(x, y) = \langle 2x + 3y, 3x \rangle$$

The largest slope at $(1, 1)$ is the magnitude of the gradient

$$|\nabla f(1, 1)| = |\langle 2(1) + 3(1), 3(1) \rangle|$$
$$= |\langle 5, 3 \rangle|$$
$$= \sqrt{25 + 9}$$
$$= \sqrt{34}$$

(b) To find the critical points we set each partial derivative equal to 0:

$$2x + 3y = 0, \quad 3x = 0$$

From the second equation $x = 0$. The first then implies $y = 0$. So the only critical point is at $(0, 0)$.

(c) We compute

$$\begin{vmatrix} \dfrac{\partial^2 f}{\partial x^2} & \dfrac{\partial^2 f}{\partial x \partial y} \\ \dfrac{\partial^2 f}{\partial y \partial x} & \dfrac{\partial^2 f}{\partial y^2} \end{vmatrix} = \begin{vmatrix} 2 & 3 \\ 3 & 0 \end{vmatrix} = -9$$

(d) Since the determinant of the matrix of second partials is negative, the critical point $(0, 0)$ is a saddle.

$$\lambda = \frac{1 \pm \sqrt{10}}{2}$$

2.(a) The magnitude of V is

$$|\langle 1, 2, 3 \rangle| = \sqrt{1^2 + 2^2 + 3^2} = \sqrt{14}$$

Hence, the desired unit vector is

$$\frac{V}{|V|} = \frac{\langle 1, 2, 3 \rangle}{\sqrt{14}} = \left\langle \frac{\sqrt{14}}{14}, \frac{\sqrt{14}}{7}, \frac{3\sqrt{14}}{14} \right\rangle$$

(b) Since $V \cdot W = |V||W|\cos\theta$, it follows that

$$
\begin{aligned}
\cos\theta &= \frac{V \cdot W}{|V||W|} \\
&= \frac{1+2+3}{\sqrt{14}\sqrt{3}} \\
&= \frac{6}{\sqrt{42}} \\
&= \frac{\sqrt{42}}{7}
\end{aligned}
$$

(c) First we find the cross product:

$$
\begin{aligned}
V \times W &= \begin{vmatrix} \mathbf{i} & \mathbf{j} & \mathbf{k} \\ 1 & 2 & 3 \\ 1 & 1 & 1 \end{vmatrix} \\
&= \mathbf{i}(2-3) - \mathbf{j}(1-3) + \mathbf{k}(1-2) \\
&= \langle -1, 2, -1 \rangle
\end{aligned}
$$

The magnitude of this vector is

$$
|\langle -1, 2, -1 \rangle| = \sqrt{1+4+1} = \sqrt{6}
$$

Hence, a unit vector that points in the desired direction is

$$
\frac{\langle -1, 2, -1 \rangle}{\sqrt{6}} = \left\langle -\frac{\sqrt{6}}{6}, \frac{\sqrt{6}}{3}, -\frac{\sqrt{6}}{6} \right\rangle
$$

Chapter 8: Calculus with Parameterizations

Problem 79

The speed is the magnitude of the derivative of the parameterization:

$$
|\langle 2t\cos t^2, -2t\sin t^2 \rangle| = \sqrt{4t^2\cos^2 t^2 + 4t^2\sin^2 t^2} = 2t
$$

Problem 80

A tangent vector is given by the derivative of the parameterization: $c'(t) = \langle 3t^2, 4t^3, 5t^4 \rangle$. The point $(1, 1, 1) = c(1)$. A tangent vector at this point is then $c'(1) = \langle 3, 4, 5 \rangle$. The magnitude of this vector is $\sqrt{9 + 16 + 25} = 5\sqrt{2}$. A unit tangent vector is thus

$$\frac{\langle 3, 4, 5 \rangle}{5\sqrt{2}} = \left\langle \frac{3\sqrt{2}}{10}, \frac{2\sqrt{2}}{5}, \frac{\sqrt{2}}{2} \right\rangle$$

Problem 81

1. The derivatives of the parameterization are

$$\frac{\partial \Psi}{\partial r} = \langle \cos \theta, \sin \theta, 2r \rangle, \text{ and}$$

$$\frac{\partial \Psi}{\partial \theta} = \langle -r \sin \theta, r \cos \theta, 0 \rangle$$

If $r = 1$ and $\theta = \frac{\pi}{6}$ these vectors are $\langle \frac{\sqrt{3}}{2}, \frac{1}{2}, 2 \rangle$ and $\langle -\frac{1}{2}, \frac{\sqrt{3}}{2}, 0 \rangle$.

2. $\begin{vmatrix} \mathbf{i} & \mathbf{j} & \mathbf{k} \\ \frac{\sqrt{3}}{2} & \frac{1}{2} & 2 \\ -\frac{1}{2} & \frac{\sqrt{3}}{2} & 0 \end{vmatrix} = \langle -\sqrt{3}, -1, 1 \rangle$

Problem 82

1. First we compute

$$\left| \frac{d\Psi}{dt} \right| = |\langle -\sin t, \cos t, 1 \rangle|$$

$$= \sqrt{\sin^2 t + \cos^2 t + 1}$$

$$= \sqrt{2}$$

Hence, if we multiply Ψ by $\frac{1}{\sqrt{2}} = \frac{\sqrt{2}}{2}$ we get the required parameterization:

$$\Phi(t) = \left(\frac{\sqrt{2}}{2} \cos t, \frac{\sqrt{2}}{2} \sin t, \frac{\sqrt{2}}{2} t \right)$$

2. First, note that

$$\frac{d\Phi}{dt} = \left\langle -\frac{\sqrt{2}}{2}\sin t, \frac{\sqrt{2}}{2}\cos t, \frac{\sqrt{2}}{2} \right\rangle$$

It now follows that

$$\frac{d^2\Phi}{dt^2} = \left\langle -\frac{\sqrt{2}}{2}\cos t, -\frac{\sqrt{2}}{2}\sin t, 0 \right\rangle$$

We now compute the dot product

$$\begin{aligned}
\frac{d\Phi}{dt} \cdot \frac{d^2\Phi}{dt^2} &= \left\langle -\frac{\sqrt{2}}{2}\sin t, \frac{\sqrt{2}}{2}\cos t, \frac{\sqrt{2}}{2} \right\rangle \cdot \left\langle -\frac{\sqrt{2}}{2}\cos t, -\frac{\sqrt{2}}{2}\sin t, 0 \right\rangle \\
&= \frac{1}{2}\sin t\cos t - \frac{1}{2}\cos t\sin t + 0 \\
&= 0
\end{aligned}$$

3. We compute the magnitude of $\frac{d^2\Phi}{dt^2}$:

$$\begin{aligned}
\left| \frac{d^2\Phi}{dt^2} \right| &= \sqrt{\left(\frac{\sqrt{2}}{2}\cos t\right)^2 + \left(\frac{\sqrt{2}}{2}\sin t\right)^2 + 0} \\
&= \sqrt{\frac{1}{2}\cos^2 t + \frac{1}{2}\sin^2 t} \\
&= \sqrt{\frac{1}{2}} \\
&= \frac{\sqrt{2}}{2}
\end{aligned}$$

Hence, if we let $N = \sqrt{2}\frac{d^2\Phi}{dt^2} = \langle -\cos t, -\sin t, 0 \rangle$ then N is a unit vector and

$$\frac{d^2\Phi}{dt^2} = \frac{\sqrt{2}}{2}N$$

So, the required κ is $\frac{\sqrt{2}}{2}$.

4. We compute

$$B = \frac{d\Phi}{dt} \times N$$

$$= \left\langle -\frac{\sqrt{2}}{2}\sin t, \frac{\sqrt{2}}{2}\cos t, \frac{\sqrt{2}}{2} \right\rangle \times \langle -\cos t, -\sin t, 0 \rangle$$

$$= \begin{vmatrix} \mathbf{i} & \mathbf{j} & \mathbf{k} \\ -\frac{\sqrt{2}}{2}\sin t & \frac{\sqrt{2}}{2}\cos t & \frac{\sqrt{2}}{2} \\ -\cos t & -\sin t & 0 \end{vmatrix}$$

$$= \left\langle \frac{\sqrt{2}}{2}\sin t, -\frac{\sqrt{2}}{2}\cos t, \frac{\sqrt{2}}{2} \right\rangle$$

We now compute the magnitude of B:

$$|B| = \sqrt{\left(\frac{\sqrt{2}}{2}\sin t\right)^2 + \left(\frac{\sqrt{2}}{2}\cos t\right)^2 + \left(\frac{\sqrt{2}}{2}\right)^2}$$

$$= \sqrt{\frac{1}{2}\sin^2 t + \frac{1}{2}\cos^2 t + \frac{1}{2}}$$

$$= 1$$

5. $\frac{dB}{dt} = \langle \frac{\sqrt{2}}{2}\cos t, \frac{\sqrt{2}}{2}\sin t, 0 \rangle$

6. Since $\frac{dB}{dt} = \langle \frac{\sqrt{2}}{2}\cos t, \frac{\sqrt{2}}{2}\sin t, 0 \rangle$ and $N = \langle -\cos t, -\sin t, 0 \rangle$ it must be that $\tau = \frac{\sqrt{2}}{2}$.

7. First, notice

$$\frac{dN}{dt} = \langle \sin t, -\cos t, 0 \rangle$$

We now do some calculation:

$$-\kappa\frac{d\Phi}{dt} + \tau B = -\frac{\sqrt{2}}{2}\left\langle -\frac{\sqrt{2}}{2}\sin t, \frac{\sqrt{2}}{2}\cos t, \frac{\sqrt{2}}{2} \right\rangle$$

$$+ \frac{\sqrt{2}}{2}\left\langle \frac{\sqrt{2}}{2}\sin t, -\frac{\sqrt{2}}{2}\cos t, \frac{\sqrt{2}}{2} \right\rangle$$

$$= \left\langle \frac{1}{2} \sin t, -\frac{1}{2} \cos t, -\frac{1}{2} \right\rangle + \left\langle \frac{1}{2} \sin t, -\frac{1}{2} \cos t, \frac{1}{2} \right\rangle$$

$$= \langle \sin t, -\cos t, 0 \rangle$$

$$= \frac{dN}{dt}$$

Problem 83

The length of $\Psi(t)$ is computed as follows:

$$\int_0^1 \left| \frac{d\Psi}{dt} \right| dt = \int_0^1 |\langle -\sin t, \cos t, 1 \rangle| \, dt$$

$$= \int_0^1 \sqrt{\cos^2 t + \sin^2 t + 1} \, dt$$

$$= \int_0^1 \sqrt{2} \, dt$$

$$= \sqrt{2} t \Big|_0^1$$

$$= \sqrt{2}$$

Problem 84

The length of $\Psi(t)$ is calculated as follows:

$$\int_0^1 \left| \frac{d\Psi}{dt} \right| dt = \int_0^1 |\langle -\sin t, \cos t, t \rangle| \, dt$$

$$= \int_0^1 \sqrt{\cos^2 t + \sin^2 t + t^2} \, dt$$

$$= \int_0^1 \sqrt{1 + t^2} \, dt$$

This is precisely the integral we obtained for the length of $(t \cos t, t \sin t)$ in Example 8-5.

Problem 85

$$\int_C f(x, y) \, d\mathbf{s} = \int_0^1 f(t^3, t)|\langle t^2, 1\rangle| \, dt$$

$$= \int_0^1 t^3 \sqrt{t^4 + 1} \, dt$$

$$= \int_1^2 \frac{1}{4} \sqrt{u} \, du$$

$$= \frac{1}{6} u^{\frac{3}{2}} \Big|_1^2$$

$$= \frac{1}{6} (2^{\frac{3}{2}} - 1)$$

Problem 86

$$\text{Area} = \int_0^1 \int_0^1 \sqrt{(2)^2 + (3)^2 + 1} \, dx \, dy$$

$$= \int_0^1 \int_0^1 \sqrt{14} \, dx \, dy$$

$$= \sqrt{14}$$

Problem 87

1. $$\text{Area} = \int_0^1 \int_0^1 \sqrt{\left(\frac{\partial f}{\partial x}\right)^2 + \left(\frac{\partial f}{\partial y}\right)^2 + 1} \, dx \, dy$$

$$= \int_0^1 \int_0^1 \sqrt{\left(\frac{x}{\sqrt{x^2 + y^2}}\right)^2 + \left(\frac{y}{\sqrt{x^2 + y^2}}\right)^2 + 1} \, dx \, dy$$

$$= \int_0^1 \int_0^1 \sqrt{2} \, dx \, dy$$

$$= \sqrt{2}$$

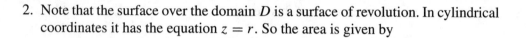

2. Note that the surface over the domain D is a surface of revolution. In cylindrical coordinates it has the equation $z = r$. So the area is given by

$$\text{Area} = 2\pi \int_0^1 r\sqrt{\left(\frac{df}{dr}\right)^2 + 1}\, dr$$

$$= 2\pi \int_0^1 r\sqrt{1^2 + 1}\, dr$$

$$= 2\sqrt{2}\pi \int_0^1 r\, dr$$

$$= \sqrt{2}\pi$$

Problem 88

A parameterization for the surface is given by

$$\Psi(\theta, \phi) = (f(\phi)\sin\phi\cos\theta,\ f(\phi)\sin\phi\sin\theta,\ f(\phi)\cos\phi)$$

The partials of this are

$$\frac{\partial\Psi}{\partial\theta} = \langle -f(\phi)\sin\phi\sin\theta,\ f(\phi)\sin\phi\cos\theta,\ 0\rangle$$

$$\frac{\partial\Psi}{\partial\phi} = \langle (f'(\phi)\sin\phi + f(\phi)\cos\phi)\cos\theta,$$

$$(f'(\phi)\sin\phi + f(\phi)\cos\phi)\sin\theta,\ f'(\phi)\cos\phi - f(\phi)\sin\phi)\rangle$$

The cross product is

$$\frac{\partial\Psi}{\partial\theta} \times \frac{\partial\Psi}{\partial\phi} = (f(\phi)f'(\phi)\sin\phi\cos\phi\cos\theta - f(\phi)^2\sin^2\phi\cos\theta)\mathbf{i}$$

$$+(f(\phi)f'(\phi)\sin\phi\cos\phi\sin\theta - f(\phi)^2\sin^2\phi\sin\theta)\mathbf{j}$$

$$-(f(\phi)f'(\phi)\sin^2\phi + f(\phi)^2\sin\phi\cos\phi)\mathbf{k}$$

With some work, the magnitude of this vector is

$$\left| \frac{\partial \Psi}{\partial \theta} \times \frac{\partial \Psi}{\partial \phi} \right| = f(\phi) \sin \phi \sqrt{(f'(\phi))^2 + f(\phi)^2}$$

And so,

$$S.A. = \int \int_0^{2\pi} f(\phi) \sin \phi \sqrt{(f'(\phi))^2 + f(\phi)^2} \, d\theta \, d\phi$$

$$= 2\pi \int f(\phi) \sin \phi \sqrt{(f'(\phi))^2 + f(\phi)^2} \, d\phi$$

Problem 89

$$S.A. = \int_{-1}^{1} \int_0^{\pi} \left| \frac{\partial \Psi}{\partial r} \times \frac{\partial \Psi}{\partial \theta} \right| \, d\theta \, dr$$

$$= \int_{-1}^{1} \int_0^{\pi} \begin{vmatrix} \mathbf{i} & \mathbf{j} & \mathbf{k} \\ \cos \theta & \sin \theta & 0 \\ -r \sin \theta & r \cos \theta & 1 \end{vmatrix} \, d\theta \, dr$$

$$= \int_{-1}^{1} \int_0^{\pi} \sqrt{1 + r^2} \, d\theta \, dr$$

$$= \pi \int_{-1}^{1} \sqrt{1 + r^2} \, dr$$

$$= \pi \sqrt{2} + \frac{\pi}{2} \ln \left(\frac{\sqrt{2} + 1}{\sqrt{2} - 1} \right) \qquad \text{(An integral table helps here.)}$$

Problem 90

The portion of the plane specified is parameterized by

$$\Psi(x, y) = (x, y, x + y)$$

$$0 \le x \le 1, \quad 0 \le y \le 1$$

We now compute

$$\int\limits_S f(x, y, z)\, dS = \int\limits_0^1 \int\limits_0^1 f(\Psi(x, y)) \left| \frac{\partial \Psi}{\partial x} \times \frac{\partial \Psi}{\partial y} \right| dx\, dy$$

$$= \int\limits_0^1 \int\limits_0^1 x + y + (x + y)\, |\langle -1, -1, 1 \rangle|\; dx\, dy$$

$$= \sqrt{3} \int\limits_0^1 \int\limits_0^1 2x + 2y\; dx\, dy$$

$$= \sqrt{3} \int\limits_0^1 x^2 + 2xy \big|_0^1 \, dy$$

$$= \sqrt{3} \int\limits_0^1 1 + 2y\; dy$$

$$= \sqrt{3}\, (y + y^2) \big|_0^1$$

$$= 2\sqrt{3}$$

Problem 91

A parameterization is given by

$$\Psi(\theta, \phi) = (\sin \phi \cos \theta, \sin \phi \sin \theta, \cos \phi)$$

$$0 \le \theta \le 2\pi, \quad 0 \le \phi \le \tfrac{\pi}{2}$$

Now calculate:

$$\int\limits_S f(x, y, z)\, dS = \int\limits_0^{\frac{\pi}{2}} \int\limits_0^{2\pi} f(\Psi(\theta, \phi)) \left| \frac{\partial \Psi}{\partial \theta} \times \frac{\partial \Psi}{\partial \phi} \right| d\theta\, d\phi$$

$$= \int\limits_0^{\frac{\pi}{2}} \int\limits_0^{2\pi} \cos \phi |\langle -\sin^2 \phi \cos \theta, -\sin^2 \phi \sin \theta, -\sin \phi \cos \phi \rangle| d\theta\, d\phi$$

$$= \int_0^{\frac{\pi}{2}} \int_0^{2\pi} \cos\phi \sin\phi \; d\theta \; d\phi$$

$$= 2\pi \int_0^{\frac{\pi}{2}} \cos\phi \sin\phi \; d\phi$$

$$= \pi \int_0^{\frac{\pi}{2}} \sin 2\phi \; d\phi$$

$$= \pi$$

Problem 92

A parameterization is given by

$$\Psi(r, \theta, z) = (r\cos\theta, r\sin\theta, z)$$

$$0 \le r \le R, \quad 0 \le \theta \le 2\pi, \quad 0 \le z \le h$$

The derivatives of this are

$$\frac{\partial\Psi}{\partial r} = \langle \cos\theta, \sin\theta, 0 \rangle$$

$$\frac{\partial\Psi}{\partial \theta} = \langle -r\sin\theta, r\cos\theta, 0 \rangle$$

$$\frac{\partial\Psi}{\partial z} = \langle 0, 0, 1 \rangle$$

We now compute volume by integrating:

$$\text{Volume} = \int_0^h \int_0^{2\pi} \int_0^R \begin{vmatrix} \cos\theta & \sin\theta & 0 \\ -r\sin\theta & r\cos\theta & 0 \\ 0 & 0 & 1 \end{vmatrix} dr \; d\theta \; dz$$

$$= \int_0^h \int_0^{2\pi} \int_0^R r \; dr \; d\theta \; dz$$

$$= \int_0^h \int_0^{2\pi} \frac{R^2}{2} \; d\theta \; dz$$

$$= \int_0^h \pi R^2 \, dz$$

$$= \pi R^2 h$$

Problem 93

First, we calculate the derivatives of the parameterization:

$$\frac{\partial \Psi}{\partial r} = \langle (1-z)\cos\theta, (1-z)\sin\theta, 0 \rangle$$

$$\frac{\partial \Psi}{\partial \theta} = \langle -(1-z)r\sin\theta, (1-z)r\cos\theta, 0 \rangle$$

$$\frac{\partial \Psi}{\partial z} = \langle -r\cos\theta, -r\sin\theta, 1 \rangle$$

Now we integrate:

$$\text{Volume} = \int_0^1 \int_0^{2\pi} \int_0^1 \left| \frac{\partial \Psi}{\partial r} \; \frac{\partial \Psi}{\partial \theta} \; \frac{\partial \Psi}{\partial z} \right| dr \, d\theta \, dz$$

$$= \int_0^1 \int_0^{2\pi} \int_0^1 \begin{vmatrix} (1-z)\cos\theta & (1-z)\sin\theta & 0 \\ -(1-z)r\sin\theta & (1-z)r\cos\theta & 0 \\ -r\cos\theta & -r\sin\theta & 1 \end{vmatrix} dr \, d\theta \, dz$$

$$= \int_0^1 \int_0^{2\pi} \int_0^1 (1-z)^2 r \, dr \, d\theta \, dz$$

$$= \int_0^1 \int_0^{2\pi} \frac{(1-z)^2}{2} \, d\theta \, dz$$

$$= \pi \int_0^1 (1-z)^2 \, dz$$

$$= \pi \left(z - z^2 + \frac{1}{3} z^3 \right) \Big|_0^1$$

$$= \frac{\pi}{3}$$

Problem 94

Letting R denote the elliptical region given by the parameterization Ψ, and $f(x, y) = 2x - y$, we have

$$\iint\limits_R 2x - y\, dx\, dy = \int_0^\pi \int_0^1 f(\Psi(r, \theta)) \left| \frac{\partial \Psi}{\partial r} \frac{\partial \Psi}{\partial \theta} \right| dr\, d\theta$$

$$= \int_0^\pi \int_0^1 [2(2r\cos\theta) - r\sin\theta] \begin{vmatrix} 2\cos\theta & \sin\theta \\ -2r\sin\theta & r\cos\theta \end{vmatrix} dr\, d\theta$$

$$= \int_0^\pi \int_0^1 (4r\cos\theta - r\sin\theta))(2r)\, dr\, d\theta$$

$$= \int_0^\pi \int_0^1 8r^2\cos\theta - 2r^2\sin\theta\, dr\, d\theta$$

$$= \int_0^\pi \frac{8}{3}r^3\cos\theta - \frac{2}{3}r^3\sin\theta \Big|_0^1 d\theta$$

$$= \int_0^\pi \frac{8}{3}\cos\theta - \frac{2}{3}\sin\theta\, d\theta$$

$$= \frac{8}{3}\sin\theta + \frac{2}{3}\cos\theta \Big|_0^\pi$$

$$= \frac{-4}{3}$$

Problem 95

$$\iint\limits_V \int \frac{1}{1 + z^2}\, dx\, dy\, dz = \int_{-2}^2 \int_0^{2\pi} \int_1^2 \frac{1}{1 + \sinh^2\omega} \left| \frac{\partial \Psi}{\partial r} \frac{\partial \Psi}{\partial \theta} \frac{\partial \Psi}{\partial \omega} \right| dr\, d\theta\, d\omega$$

$$= \int_{-2}^2 \int_0^{2\pi} \int_1^2 \frac{1}{\cosh^2\omega} \begin{vmatrix} \cosh\omega\cos\theta & \cosh\omega\sin\theta & 0 \\ -r\cosh\omega\sin\theta & r\cosh\omega\cos\theta & 0 \\ r\sinh\omega\cos\theta & r\sinh\omega\sin\theta & \cosh\omega \end{vmatrix} dr\, d\theta\, d\omega$$

$$= \int\limits_{-2}^{2} \int\limits_{0}^{2\pi} \int\limits_{1}^{2} \frac{r\cosh^3\omega}{\cosh^2\omega}\, dr\, d\theta\, d\omega$$

$$= \int\limits_{-2}^{2} \int\limits_{0}^{2\pi} \int\limits_{1}^{2} r\cosh\omega\, dr\, d\theta\, d\omega$$

$$= \int\limits_{-2}^{2} \int\limits_{0}^{2\pi} \frac{r^2}{2}\cosh\omega \Big|_{1}^{2}\, d\theta\, d\omega$$

$$= \frac{3}{2} \int\limits_{-2}^{2} \int\limits_{0}^{2\pi} \cosh\omega\, d\theta\, d\omega$$

$$= 3\pi \int\limits_{-2}^{2} \cosh\omega\, d\omega$$

$$= 3\pi \,\sinh\omega\,|^{2}_{-2}$$

$$= 6\pi \sinh 2$$

Problem 96

The region V is easily parameterized using cylindrical coordinates:

$$Psi(r, \theta, z) = (r\cos\theta, r\sin\theta, z)$$

$$0 \le r \le 1, \quad 0 \le \theta \le 2\pi, \quad 0 \le z \le 2$$

To use this parameterization to evaluate the integral, we will need the determinant of the matrix of partial derivatives of *Psi*:

$$\begin{vmatrix} \cos\theta & \sin\theta & 0 \\ -r\sin\theta & r\cos\theta & 0 \\ 0 & 0 & 1 \end{vmatrix} = r$$

We now integrate

$$\int_V z \, dx \, dy \, dz = \int_0^2 \int_0^{2\pi} \int_0^1 z(r) \, dr \, d\theta \, dz$$

$$= \int_0^2 \int_0^{2\pi} \frac{1}{2} z \, d\theta \, dz$$

$$= \int_0^2 \pi z \, dz$$

$$= 2\pi$$

Problem 97

The region R is easily parameterized with polar coordinates:

$$\Psi(r, \theta) = (r \cos \theta, r \sin \theta)$$

$$0 \leq r \leq 2, \quad 0 \leq \theta \leq \frac{\pi}{4}$$

The determinant of the matrix of partials of Ψ is r. Hence,

$$\int_R 1 + \frac{y^2}{x^2} \, dx \, dy = \int_0^{\frac{\pi}{4}} \int_0^2 \left(1 + \frac{(r \sin \theta)^2}{(r \cos \theta)^2}\right) (r) \, dr \, d\theta$$

$$= \int_0^{\frac{\pi}{4}} \int_0^2 \left(1 + \tan^2 \theta\right) (r) \, dr \, d\theta$$

$$= \int_0^{\frac{\pi}{4}} \int_0^2 r \sec^2 \theta \, dr \, d\theta$$

$$= \int_0^{\frac{\pi}{4}} 2 \sec^2 \theta \, d\theta$$

$$= 2 \tan \theta |_0^{\frac{\pi}{4}}$$

$$= 2$$

Chapter 8 Quiz

Problem 98

1.(a) $\Psi(y) = (\sin y, y)$

(b) The point $(0, 0)$ is $\Psi(0)$. So a tangent vector is given by the derivative of the parameterization, $\frac{d\Psi}{dt} = (\cos y, 1)$, at $y = 0$. This is the vector $(1, 1)$. A unit tangent vector is thus

$$\frac{(1, 1)}{|(1, 1)|} = \frac{(1, 1)}{\sqrt{2}} = \left\langle \frac{\sqrt{2}}{2}, \frac{\sqrt{2}}{2} \right\rangle$$

2.(a)
$$\Psi(x, y, t) = (x, y, t(x^2 + y^2))$$
$$0 \le x, y, t \le 1$$

(b) We will need the partials of Ψ:

$$\frac{\partial \Psi}{\partial x} = (1, 0, 2xt)$$

$$\frac{\partial \Psi}{\partial y} = (0, 1, 2yt)$$

$$\frac{\partial \Psi}{\partial t} = (0, 0, x^2 + y^2)$$

Next we must compute the determinant of the matrix of partials:

$$\begin{vmatrix} 1 & 0 & 2xt \\ 0 & 1 & 2yt \\ 0 & 0 & x^2 + y^2 \end{vmatrix} = x^2 + y^2$$

Finally, we integrate

$$\iiint_Q \frac{z}{x^2 + y^2} \, dx \, dy \, dz = \int_0^1 \int_0^1 \int_0^1 \frac{t(x^2 + y^2)}{x^2 + y^2}(x^2 + y^2) \, dx \, dy \, dt$$

$$= \int_0^1 \int_0^1 \int_0^1 tx^2 + ty^2 \, dx \, dy \, dt$$

$$= \int_0^1 \int_0^1 \frac{1}{3}t + ty^2 \, dy \, dt$$

$$= \int_0^1 \frac{1}{3}t + \frac{1}{3}t \, dt$$

$$= \frac{1}{3}$$

3. First, we compute the partials of ϕ:

$$\frac{\partial \phi}{\partial r} = \langle \cosh t, \sinh t \rangle$$

$$\frac{\partial \phi}{\partial t} = \langle r \sinh t, r \cosh t \rangle$$

Next we will need the determinant of the matrix of partials:

$$\begin{vmatrix} \cosh t & \sinh t \\ r \sinh t & r \cosh t \end{vmatrix} = r$$

Now we integrate

$$\iint_R x^2 - y^2 \, dx \, dy = \int_{-1}^1 \int_0^1 ((r \cosh t)^2 - (r \sinh t)^2)(r) \, dr \, dt$$

$$= \int_{-1}^1 \int_0^1 r^3 \, dr \, dt$$

$$= \int_{-1}^1 \frac{1}{4} \, dt$$

$$= \frac{1}{2}$$

Chapter 9: Vector Fields and Derivatives

Problem 99

1.

2.

3.

Problem 100

1. $\langle 1, 1 \rangle$ 2. $\langle 1, z, y \rangle$ 3. $\langle y + z, x + z, x + y \rangle$

Problem 101

1. $\dfrac{\partial}{\partial x}(y) + \dfrac{\partial}{\partial y}(z) + \dfrac{\partial}{\partial z}(x) = 0 + 0 + 0 = 0$

2. $\dfrac{\partial}{\partial x}(x + y) + \dfrac{\partial}{\partial y}(x - y) + \dfrac{\partial}{\partial z}(z) = 1 - 1 + 1 = 1$

3. $\dfrac{\partial}{\partial x}(x^2 + y^2) + \dfrac{\partial}{\partial y}(x^2 - y^2) + \dfrac{\partial}{\partial z}(z^2) = 2x - 2y + 2z$

Problem 102

$$\nabla \cdot \nabla f = \nabla \cdot \left\langle \frac{\partial f}{\partial x}, \frac{\partial f}{\partial y}, \frac{\partial f}{\partial z} \right\rangle$$

$$= \frac{\partial^2 f}{\partial x^2} + \frac{\partial^2 f}{\partial y^2} + \frac{\partial^2 f}{\partial z^2}$$

Problem 103

We let $\mathbf{F} = \langle f_x, f_y, f_z \rangle$ and $\mathbf{G} = \langle g_x, g_y, g_z \rangle$. Then

$$\mathbf{F} \times \mathbf{G} = \langle f_y g_z - f_z g_y, f_z g_x - f_x g_z, f_x g_y - f_y g_x \rangle$$

and so

$$\nabla \cdot (\mathbf{F} \times \mathbf{G}) = \frac{\partial}{\partial x}(f_y g_z - f_z g_y) + \frac{\partial}{\partial y}(f_z g_x - f_x g_z)$$

$$+ \frac{\partial}{\partial z}(f_x g_y - f_y g_x)$$

$$= \frac{\partial f_y}{\partial x} g_z + f_y \frac{\partial g_z}{\partial x} - \frac{\partial f_z}{\partial x} g_y - f_z \frac{\partial g_y}{\partial x}$$

$$+ \frac{\partial f_z}{\partial y} g_x + f_z \frac{\partial g_x}{\partial y} - \frac{\partial f_x}{\partial y} g_z - f_x \frac{\partial g_z}{\partial y}$$

$$+ \frac{\partial f_x}{\partial z} g_y + f_x \frac{\partial g_y}{\partial z} - \frac{\partial f_y}{\partial z} g_x - f_y \frac{\partial g_x}{\partial z}$$

$$= \langle g_x, g_y, g_z \rangle \cdot \left\langle \frac{\partial f_z}{\partial y} - \frac{\partial f_y}{\partial z}, \frac{\partial f_x}{\partial z} - \frac{\partial f_z}{\partial x}, \frac{\partial f_y}{\partial x} - \frac{\partial f_x}{\partial y} \right\rangle$$

$$- \langle f_x, f_y, f_z \rangle \cdot \left\langle \frac{\partial g_z}{\partial y} - \frac{\partial g_y}{\partial z}, \frac{\partial g_x}{\partial z} - \frac{\partial g_z}{\partial x}, \frac{\partial g_y}{\partial x} - \frac{\partial g_x}{\partial y} \right\rangle$$

$$= \mathbf{G} \cdot (\nabla \times \mathbf{F}) - \mathbf{F} \cdot (\nabla \times \mathbf{G})$$

Problem 104

1. $\operatorname{curl} \mathbf{V} = \nabla \times \mathbf{V}$

$$= \begin{vmatrix} \mathbf{i} & \mathbf{j} & \mathbf{k} \\ \frac{\partial}{\partial x} & \frac{\partial}{\partial y} & \frac{\partial}{\partial z} \\ x+y & x-z & y+z \end{vmatrix}$$

$$= \langle 2, 0, 0 \rangle$$

2. $\langle 2x, 0, -2z \rangle$

Problem 105

1. $\nabla \times (\nabla f) = \nabla \times \left\langle \frac{\partial f}{\partial x}, \frac{\partial f}{\partial y}, \frac{\partial f}{\partial z} \right\rangle$

$$= \begin{vmatrix} \mathbf{i} & \mathbf{j} & \mathbf{k} \\ \frac{\partial}{\partial x} & \frac{\partial}{\partial y} & \frac{\partial}{\partial z} \\ \frac{\partial f}{\partial x} & \frac{\partial f}{\partial y} & \frac{\partial f}{\partial z} \end{vmatrix}$$

$$= \left\langle \frac{\partial^2 f}{\partial y \partial z} - \frac{\partial^2 f}{\partial z \partial y}, \frac{\partial^2 f}{\partial z \partial x} - \frac{\partial^2 f}{\partial x \partial z}, \frac{\partial^2 f}{\partial x \partial y} - \frac{\partial^2 f}{\partial y \partial x} \right\rangle$$

$$= \langle 0, 0, 0 \rangle$$

2. $\nabla \cdot (\nabla \times \mathbf{V}) = \nabla \cdot \left\langle \frac{\partial h}{\partial y} - \frac{\partial g}{\partial z}, \frac{\partial f}{\partial z} - \frac{\partial h}{\partial x}, \frac{\partial g}{\partial x} - \frac{\partial f}{\partial y} \right\rangle$

$$= \frac{\partial}{\partial x} \left(\frac{\partial h}{\partial y} - \frac{\partial g}{\partial z} \right) - \frac{\partial}{\partial y} \left(\frac{\partial h}{\partial x} - \frac{\partial f}{\partial z} \right) + \frac{\partial}{\partial z} \left(\frac{\partial g}{\partial x} - \frac{\partial f}{\partial y} \right)$$

$$= \frac{\partial^2 h}{\partial x \partial y} - \frac{\partial^2 g}{\partial x \partial z} - \frac{\partial^2 h}{\partial y \partial x} + \frac{\partial^2 f}{\partial y \partial z} + \frac{\partial^2 g}{\partial z \partial x} - \frac{\partial^2 f}{\partial z \partial y}$$

$$= 0$$

Chapter 9 Quiz

Problem 106

1.
$$\begin{vmatrix} \mathbf{i} & \mathbf{j} & \mathbf{k} \\ \dfrac{\partial}{\partial x} & \dfrac{\partial}{\partial y} & \dfrac{\partial}{\partial z} \\ \dfrac{-y}{x^2+y^2} & \dfrac{x}{x^2+y^2} & 0 \end{vmatrix} = \langle 0, 0, 0 \rangle$$

2. $\dfrac{\partial}{\partial x}(x^2 + y^2) + \dfrac{\partial}{\partial y}(y^2 - x^2) + 0 = 2x + 2y$

3. $\langle 2x\sin(y-z), x^2\cos(y-z), -x^2\cos(y-z) \rangle$

4. Suppose $\mathbf{W} = \langle f, g, h \rangle$. Then the last component of $\nabla \times \mathbf{W}$ is $\frac{\partial g}{\partial x} - \frac{\partial f}{\partial y}$. This can equal $\frac{y}{x}$ if either $\frac{\partial g}{\partial x} = \frac{y}{x}$ or $-\frac{\partial f}{\partial y} = \frac{y}{x}$. In the former case $g(x, y, z) = y\ln x$. In the latter case $f(x, y, x) = -\frac{1}{x}$. So two vector fields that have the required curl are $\langle 0, y\ln x, 0 \rangle$ and $\langle -\frac{1}{x}, 0, 0 \rangle$.

Chapter 10: Integrating Vector Fields

Problem 107

The circle is parameterized by

$$\Psi(t) = (\cos t, \sin t), \quad 0 \le t \le 2\pi$$

Note that this parameterization agrees with the specified orientation. Now we may integrate

$$\int_C \mathbf{W} \cdot d\mathbf{s} = \int_0^{2\pi} \mathbf{W}(\Psi(t)) \cdot \frac{d\Psi}{dt}\, dt$$

$$= \int_0^{2\pi} \langle -\sin t, \cos t \rangle \cdot \langle -\sin t, \cos t \rangle\, dt$$

$$= \int_0^{2\pi} \sin^2 t + \cos^2 t\, dt$$

$$= \int_0^{2\pi} dt$$

$$= 2\pi$$

Problem 108

$$\int_C \mathbf{W} \cdot d\mathbf{s} = \int_0^1 \mathbf{W}(\Psi(t)) \cdot \frac{d\Psi}{dt} \, dt$$

$$= \int_0^1 \langle t^3, t^2(1-t)^2, t+1-t \rangle \cdot \langle 2t, 1, -1 \rangle \, dt$$

$$= \int_0^1 3t^4 - 2t^3 + t^2 - 1 \, dt$$

$$= \frac{3}{5}t^5 - \frac{1}{2}t^4 + \frac{1}{3}t^3 - t \Big|_0^1$$

$$= \frac{3}{5} - \frac{1}{2} + \frac{1}{3} - 1$$

$$= -\frac{17}{30}$$

Problem 109

1. First, note that $\nabla f = \langle y^2, 2xy \rangle$. Now we integrate

$$\int_C \nabla f \cdot d\mathbf{s} = \int_{-1}^2 \nabla f(\Psi(t)) \cdot \frac{d\Psi}{dt} \, dt$$

$$= \int_{-1}^2 \langle t^4, 2t^3 \rangle \cdot \langle 1, 2t \rangle \, dt$$

$$= \int_{-1}^2 5t^4 \, dt$$

$$= t^5\big|_{-1}^{2}$$
$$= 32 - (-1)$$
$$= 33$$

2. $f(\Psi(2)) - f(\Psi(-1)) = f(2, 4) - f(-1, 1)$
$$= 32 - (-1)$$
$$= 33$$

Problem 110

The curve C is parameterized by

$$\Psi(t) = (t, t^2), \quad 0 \le t \le 1$$

The derivative of this parameterization is $\langle 1, 2t \rangle$.

$$\int_C \mathbf{W} \cdot d\mathbf{s} = \int_0^1 \langle -t^4, t^3 \rangle \cdot \langle 1, 2t \rangle \, dt$$

$$= \int_0^1 t^4 \, dt$$

$$= \frac{1}{5}$$

Problem 111

The unit sphere is parameterized with spherical coordinates by

$$\Psi(\theta, \phi) = (\sin \phi \cos \theta, \sin \phi \sin \theta, \cos \phi)$$
$$0 \le \theta \le 2\pi, \quad 0 \le \phi \le \pi$$

The derivatives of this are

$$\frac{\partial \Psi}{\partial \theta} = \langle -\sin \phi \sin \theta, \sin \phi \cos \theta, 0 \rangle$$

$$\frac{\partial \Psi}{\partial \theta} = \langle \cos \phi \cos \theta, \cos \phi \sin \theta, -\sin \phi \rangle$$

And so,

$$\frac{\partial \Psi}{\partial \theta} \times \frac{\partial \Psi}{\partial \phi} = \begin{vmatrix} \mathbf{i} & \mathbf{j} & \mathbf{k} \\ -\sin \phi \sin \theta & \sin \phi \cos \theta & 0 \\ \cos \phi \cos \theta & \cos \phi \sin \theta & -\sin \phi \end{vmatrix}$$

$$= \langle -\sin^2 \phi \cos \theta, -\sin^2 \phi \sin \theta, -\sin \phi \cos \phi \rangle$$

Notice that at the point $\Psi(0, \frac{\pi}{2}) = (1, 0, 0)$ the vector $\frac{\partial \Psi}{\partial \theta} \times \frac{\partial \Psi}{\partial \phi}$ is $\langle 1, 0, 0 \rangle$, which agrees with the specified orientation. We may thus integrate to find the correct answer:

$$\int_S \mathbf{W} \cdot d\mathbf{S} = \int_0^\pi \int_0^{2\pi} \mathbf{W}(\Psi(\theta, \phi)) \cdot \left(\frac{\partial \Psi}{\partial \theta} \times \frac{\partial \Psi}{\partial \phi} \right) d\theta \, d\phi$$

$$= \int_0^\pi \int_0^{2\pi} \langle \sin \phi \cos \theta, \sin \phi \sin \theta, \cos \phi \rangle \cdot \left(\frac{\partial \Psi}{\partial \theta} \times \frac{\partial \Psi}{\partial \phi} \right) d\theta \, d\phi$$

$$= \int_0^\pi \int_0^{2\pi} -\sin \phi \, d\theta \, d\phi$$

$$= \int_0^\pi -2\pi \sin \phi \, d\phi$$

$$= 2\pi \cos \phi |_0^\pi$$

$$= -4\pi$$

Problem 112

First, we calculate the partials of the parameterization:

$$\frac{\partial \Psi}{\partial u} = \langle 1, 0, 2u \rangle$$

$$\frac{\partial \Psi}{\partial v} = \langle 0, 1, 2v \rangle$$

Now, the cross product is given by

$$\frac{\partial \Psi}{\partial u} \times \frac{\partial \Psi}{\partial v} = \begin{vmatrix} \mathbf{i} & \mathbf{j} & \mathbf{k} \\ 1 & 0 & 2u \\ 0 & 1 & 2v \end{vmatrix} = \langle -2u, -2v, 1 \rangle$$

Finally, we integrate

$$\int_S \mathbf{W} \cdot d\mathbf{S} = \int_0^1 \int_0^1 \mathbf{W}(\Psi(u, v)) \cdot \left(\frac{\partial \Psi}{\partial u} \times \frac{\partial \Psi}{\partial v} \right) \, du \, dv$$

$$= \int_0^1 \int_0^1 \langle u(u^2 + v^2), v(u^2 + v^2), 0 \rangle \cdot \langle -2u, -2v, 1 \rangle \, du \, dv$$

$$= \int_0^1 \int_0^1 -2u^2(u^2 + v^2) - 2v^2(u^2 + v^2) \, du \, dv$$

$$= \int_0^1 \int_0^1 -2u^4 - 4u^2v^2 - 2v^4 \, du \, dv$$

$$= \int_0^1 -\frac{2}{5} - \frac{4}{3}v^2 - 2v^4 \, dv$$

$$= -\frac{2}{5} - \frac{4}{9} - \frac{2}{5}$$

$$= -\frac{56}{45}$$

Problem 113

First, we calculate the partials of the parameterization:

$$\frac{\partial \Psi}{\partial \theta} = \langle -\cos \phi \sin \theta, \cos \phi \cos \theta, 0 \rangle$$

$$\frac{\partial \Psi}{\partial \phi} = \langle -\sin \phi \cos \theta, -\sin \phi \sin \theta, \cos \phi \rangle$$

Now, the cross product of these vectors is

$$\frac{\partial \Psi}{\partial \theta} \times \frac{\partial \Psi}{\partial \phi} = \begin{vmatrix} \mathbf{i} & \mathbf{j} & \mathbf{k} \\ -\cos \phi \sin \theta & \cos \phi \cos \theta & 0 \\ -\sin \phi \cos \theta & -\sin \phi \sin \theta & \cos \phi \end{vmatrix}$$

$$= \langle \cos^2 \phi \cos \theta, \cos^2 \phi \sin \theta, \cos \phi \sin \phi \rangle$$

We are now ready to integrate

$$\int_S \mathbf{W} \cdot d\mathbf{S} = \int_{-\frac{\pi}{4}}^{\frac{\pi}{4}} \int_0^{2\pi} \left\langle \frac{1}{\cos \phi \cos \theta}, \frac{1}{\cos \phi \sin \theta}, 0 \right\rangle$$

$$\cdot \langle \cos^2 \phi \cos \theta, \cos^2 \phi \sin \theta, \cos \phi \sin \phi \rangle \, d\theta \, d\phi$$

$$= \int_{-\frac{\pi}{4}}^{\frac{\pi}{4}} \int_0^{2\pi} 2 \cos \phi \, d\theta \, d\phi$$

$$= \int_{-\frac{\pi}{4}}^{\frac{\pi}{4}} 4\pi \cos \phi \, d\phi$$

$$= 4\pi \sin \phi \big|_{-\frac{\pi}{4}}^{\frac{\pi}{4}}$$

$$= 4\pi \sqrt{2}$$

Problem 114

First, we compute $\nabla \times \mathbf{W}$:

$$\begin{vmatrix} \mathbf{i} & \mathbf{j} & \mathbf{k} \\ \frac{\partial}{\partial x} & \frac{\partial}{\partial y} & \frac{\partial}{\partial z} \\ xz & yz & 0 \end{vmatrix} = \langle y, x, 0 \rangle$$

A parameterization for S is given (via spherical coordinates) by

$$\Psi(\theta, \phi) = (\sin \phi \cos \theta, \sin \phi \sin \theta, \cos \phi)$$

$$0 \leq \theta \leq \frac{\pi}{4}, \quad 0 \leq \phi \leq \frac{\pi}{2}$$

The partials of this are

$$\frac{\partial \Psi}{\partial \theta} = \langle -\sin \phi \sin \theta, \sin \phi \cos \theta, 0 \rangle$$

$$\frac{\partial \Psi}{\partial \theta} = \langle \cos \phi \cos \theta, \cos \phi \sin \theta, -\sin \phi \rangle$$

And so,

$$\frac{\partial \Psi}{\partial \theta} \times \frac{\partial \Psi}{\partial \phi} = \begin{vmatrix} \mathbf{i} & \mathbf{j} & \mathbf{k} \\ -\sin \phi \sin \theta & \sin \phi \cos \theta & 0 \\ \cos \phi \cos \theta & \cos \phi \sin \theta & -\sin \phi \end{vmatrix}$$

$$= \langle -\sin^2 \phi \cos \theta, -\sin^2 \phi \sin \theta, -\sin \phi \cos \phi \rangle$$

Now notice that at the point $\Psi(\frac{\pi}{2}, \frac{\pi}{2}) = (0, 1, 0)$ the vector $\frac{\partial \Psi}{\partial \theta} \times \frac{\partial \Psi}{\partial \phi}$ is equal to $\langle 0, 1, 0 \rangle$, which disagrees with the specified orientation. We must therefore remember to switch the sign of our final answer. Finally, we can integrate

$$\int_S (\nabla \times \mathbf{W}) \cdot d\mathbf{S} = \int_0^{\frac{\pi}{2}} \int_0^{\frac{\pi}{4}} \nabla \times \mathbf{W}(\Psi(\theta, \phi)) \cdot \left(\frac{\partial \Psi}{\partial \theta} \times \frac{\partial \Psi}{\partial \phi} \right) d\theta \, d\phi$$

$$= \int_0^{\frac{\pi}{2}} \int_0^{\frac{\pi}{4}} \langle \sin \phi \sin \theta, \sin \phi \cos \theta, 0 \rangle \cdot \left(\frac{\partial \Psi}{\partial \theta} \times \frac{\partial \Psi}{\partial \phi} \right) d\theta \, d\phi$$

$$= \int_0^{\frac{\pi}{2}} \int_0^{\frac{\pi}{4}} -2 \sin^3 \phi \sin \theta \cos \theta \, d\theta \, d\phi$$

$$= \int_0^{\frac{\pi}{2}} \int_0^{\frac{\pi}{4}} -\sin^3 \phi \sin 2\theta \, d\theta \, d\phi$$

$$= \int_0^{\frac{\pi}{2}} \frac{1}{2} \sin^3 \phi \cos 2\theta \Big|_0^{\frac{\pi}{4}} \, d\phi$$

$$= \int_{0}^{\frac{\pi}{2}} -\frac{1}{2}(1 - \cos^2 \phi) \sin \phi \, d\phi$$

$$= \frac{1}{2}(\cos \phi - \frac{1}{3} \cos^3 \phi)\Big|_{0}^{\frac{\pi}{2}}$$

$$= -\frac{1}{3}$$

The correct answer is therefore $\frac{1}{3}$.

Chapter 10 Quiz

Problem 115

1. A parameterization for C is given by

$$\Psi(t) = (t^2, t), \quad 0 \le t \le 1$$

The derivative of this is

$$\frac{d\Psi}{dt} = \langle 2t, 1 \rangle$$

Note that at the point $\Psi(0) = (0, 0)$ this vector is $\langle 0, 1 \rangle$, which disagrees with the orientation indicated in the figure. Thus, we will have to negate our final answer. We now integrate

$$\int_{C} \langle 1, 1 \rangle \cdot d\mathbf{s} = -\int_{0}^{1} \langle 1, 1 \rangle \cdot \langle 2t, 1 \rangle \, dt$$

$$= -\int_{0}^{1} 2t + 1 \, dt$$

$$= -2$$

2. The partials of Ψ are given by

$$\frac{\partial \Psi}{\partial r} = \langle \cos \theta, \sin \theta, 1 \rangle$$

$$\frac{\partial \Psi}{\partial \theta} = \langle -r \sin \theta, r \cos \theta, 0 \rangle$$

The cross product of these vectors is

$$\frac{\partial \Psi}{\partial r} \times \frac{\partial \Psi}{\partial \theta} = \begin{vmatrix} \mathbf{i} & \mathbf{j} & \mathbf{k} \\ \cos \theta & \sin \theta & 1 \\ -r \sin \theta & r \cos \theta & 0 \end{vmatrix} = \langle -r \cos \theta, -r \sin \theta, r \rangle$$

We are now prepared to integrate

$$\int_S \left\langle \frac{1}{x}, -\frac{1}{y}, z \right\rangle \cdot d\mathbf{S} = \int_0^{2\pi} \int_1^2 \left\langle \frac{1}{r \cos \theta}, \frac{-1}{r \sin \theta}, r \right\rangle \cdot \langle -r \cos \theta, -r \sin \theta, r \rangle \ dr \ d\theta$$

$$= \int_0^{2\pi} \int_1^2 -1 + 1 + r^2 \ dr \ d\theta$$

$$= \int_0^{2\pi} \frac{1}{3} r^3 \bigg|_1^2 \ d\theta$$

$$= \int_0^{2\pi} \frac{7}{3} \ d\theta$$

$$= \frac{14\pi}{3}$$

Chapter 11: Integration Theorems

Problem 116

Notice that $\mathbf{W} = \langle 1, 1, 1 \rangle = \nabla(x + y + z)$. Hence, if we let $f(x, y, z) = x + y + z$ then

$$\int_C \mathbf{W} \cdot d\mathbf{s} = f(1, 1, 1) - f(1, 0, 1) = 3 - 2 = 1$$

Problem 117

The trick is to notice that $\mathbf{W} = \nabla f$, where $f(x, y) = x \sin y$. Then

$$\int_C \mathbf{W} \cdot d\mathbf{s} = f\left(2, \frac{\pi}{2}\right) - f(0, 0)$$

$$= 2 \sin \frac{\pi}{2} - 0 \sin 0$$

$$= 2$$

Problem 118

First, notice that $f(x, y) = x^2 - 4x + 4 + y^2 + 2y + 1 = (x - 2)^2 + (y + 1)^2$. Thus the graph of $f(x, y)$ can be obtained from the graph of $z = x^2 + y^2$ (a paraboloid) by shifting 2 units in the positive x-direction and 1 unit in the negative y-direction. This puts the "bottom" of the paraboloid at the point $(2, -1)$. Hence, if (a, b) is any point other than $(2, -1)$ then $f(a, b) > 0$. The result now follows easily:

$$\int_C \nabla f \cdot d\mathbf{s} = f(a, b) - f(2, -1) = f(a, b) > 0$$

Problem 119

Suppose the beginning and ending point of C is p. Then

$$\int_C \nabla f \cdot d\mathbf{s} = f(p) - f(p) = 0$$

Problem 120

Suppose $f(x)$ is a function. Let Ψ be a parameterization of $[a, b]$ given by

$$\Psi(x) = x, \quad a \leq x \leq b$$

By the theorem,

$$\int_C \nabla f \cdot d\mathbf{s} = f(b) - f(a)$$

We now follow the definitions:

$$\int_C \nabla f \cdot d\mathbf{s} = \int_a^b \langle f'(x) \rangle \cdot \frac{d\Psi}{dx}\, dx$$

$$= \int_a^b \langle f'(x) \rangle \cdot \langle 1 \rangle\ dx$$

$$= \int_a^b f'(x)\, dx$$

Putting this all together gives

$$\int_a^b f'(x)\, dx = f(b) - f(a)$$

Problem 121

When you are coming down the mountain the force of gravity is working *with* you, not *against* you. So the work done against gravity during this portion of the trip is negative. This cancels out the extra positive work you did to get to the top of the mountain.

Problem 122

$$\int_{\partial Q} \langle -y^2, x^2 \rangle \cdot d\mathbf{s} = \int_0^1 \int_0^1 \frac{\partial}{\partial x} x^2 - \frac{\partial}{\partial y}(-y^2)\, dx\, dy$$

$$= \int_0^1 \int_0^1 2x + 2y\, dx\, dy$$

$$= \int_0^1 1 + 2y\, dy$$

$$= 2$$

Problem 123

Let D be the horizontal line segment that connects $(0, 0)$ to $(a, 0)$. Then $C \cup D$ bounds a rectangular region Q. Green's Theorem says

$$\int_C \langle y, x \rangle \cdot d\mathbf{s} + \int_D \langle y, x \rangle \cdot d\mathbf{s} = \int_{C \cup D} \langle y, x \rangle \cdot d\mathbf{s}$$

$$= \int\int_Q \frac{\partial}{\partial x} x - \frac{\partial}{\partial y} y \, dx \, dy$$

$$= 0$$

We conclude

$$\int_C \langle y, x \rangle \cdot d\mathbf{s} = - \int_D \langle y, x \rangle \cdot d\mathbf{s}$$

and so the answer only depends on a.

Problem 124

By Green's Theorem,

$$\int_{\partial \sigma} \mathbf{W} \cdot d\mathbf{s} = \int_\sigma \frac{\partial}{\partial x}(x^2) - \frac{\partial}{\partial y}(-y^2) \, dx \, dy$$

$$= \int_\sigma 2x + 2y \, dx \, dy$$

To evaluate this integral we will need the determinant of the matrix of partial derivatives of ϕ:

$$\begin{vmatrix} 2 & 1 \\ -1 & 1 \end{vmatrix} = 3$$

We now integrate

$$\int_{\sigma} 2x + 2y\ dx\ dy = \int_0^1 \int_1^2 (2(2u-v) + 2(u+v))\ 3\ du\ dv$$

$$= \int_0^1 \int_1^2 18u\ du\ dv$$

$$= \int_0^1 27\ dv$$

$$= 27$$

Problem 125

The area of any region Q in $\mathbb{R}^2$ can be calculated by the integral

$$\iint_Q dx\ dy$$

Green's Theorem says that this will be equal to the integral of $\mathbf{W} = \langle f(x), g(x) \rangle$ around ∂Q if

$$\frac{\partial g}{\partial x} - \frac{\partial f}{\partial y} = 1$$

One suitable choice for $\mathbf{W}$ would thus be $\langle 0, x \rangle$. We now integrate this vector field over the unit circle, by using the usual parameterization

$$\Psi(t) = (\cos t, \sin t), \quad 0 \le t \le 2\pi$$

$$\int_{\partial Q} \mathbf{W} \cdot d\mathbf{s} = \int_0^{2\pi} \langle 0, \cos t \rangle \cdot \langle -\sin t, \cos t \rangle\ dt$$

$$= \int_0^{2\pi} \cos^2 t\ dt$$

$$= \frac{1}{2}t + \frac{1}{4}\sin 2t \Big|_0^{2\pi}$$

$$= \pi$$

Problem 126

1. Let C_1 and C_2 be two circles centered on the origin, oriented counterclockwise. Let Q be the region between these circles. Then ∂Q has two pieces. One is C_1, and the other is C_2 with the opposite orientation. As a shorthand only we write this as $C_1 - C_2$. Green's Theorem now says

$$0 = \int\int_Q 0 \, dx \, dy$$

$$= \int_{C_1 - C_2} \langle f(x), g(x) \rangle \cdot d\mathbf{s}$$

$$= \int_{C_1} \langle f(x), g(x) \rangle \cdot d\mathbf{s} - \int_{C_2} \langle f(x), g(x) \rangle \cdot d\mathbf{s}$$

And hence, $\int_{C_1} \langle f(x), g(x) \rangle \cdot d\mathbf{s} = \int_{C_2} \langle f(x), g(x) \rangle \cdot d\mathbf{s}$.

2. Using the previous part, we just have to show $\frac{\partial g}{\partial x} - \frac{\partial f}{\partial y} = 0$.

$$\frac{\partial}{\partial x} \frac{x}{x^2 + y^2} - \frac{\partial}{\partial y} \frac{-y}{x^2 + y^2}$$

$$= \left(\frac{1}{x^2 + y^2} - \frac{2x^2}{(x^2 + y^2)^2} \right) - \left(\frac{-1}{x^2 + y^2} + \frac{2y^2}{(x^2 + y^2)^2} \right)$$

$$= \frac{2}{x^2 + y^2} - \frac{2x^2 + 2y^2}{(x^2 + y^2)^2}$$

$$= 0$$

3. Note that on the unit circle $x^2 + y^2 = 1$, so, on this circle the integral of $\mathbf{W}$ is the same as the integral of $\langle -y, x \rangle$.

$$\int_C \mathbf{W} \cdot d\mathbf{s} = \int_C \langle -y, x \rangle \cdot d\mathbf{s} = \int_0^{2\pi} \langle -\sin t, \cos t \rangle \cdot \langle -\sin t, \cos t \rangle \ dt$$

$$= \int_0^{2\pi} \sin^2 t + \cos^2 t \ dt$$

$$= \int_0^{2\pi} dt$$

$$= 2\pi$$

Problem 127

1. By Green's Theorem,

$$\int_{\partial \sigma} \mathbf{W} \cdot d\mathbf{s} = \int_\sigma \frac{\partial}{\partial x}(e^y) - \frac{\partial}{\partial y}(x^2) \ dx \ dy = 0$$

2. The curve C is parameterized by

$$\Psi(t) = (t, 0), \quad -1 \leq t \leq 1$$

The derivative of this parameterization is $\langle 1, 0 \rangle$, so

$$\int_C \mathbf{W} \cdot d\mathbf{s} = \int_{-1}^{1} \langle t^2, e^0 \rangle \cdot \langle 1, 0 \rangle \ dt$$

$$= \int_{-1}^{1} t^2 \ dt$$

$$= \frac{2}{3}$$

3. Let D be the top half of the unit circle (oriented counterclockwise). Then $C \cup D = \partial \sigma$. So we have

$$
\begin{aligned}
0 &= \int_{\partial \sigma} \mathbf{W} \cdot d\mathbf{s} \\
&= \int_{C \cup D} \mathbf{W} \cdot d\mathbf{s} \\
&= \int_{C} \mathbf{W} \cdot d\mathbf{s} + \int_{D} \mathbf{W} \cdot d\mathbf{s} \\
&= \frac{2}{3} + \int_{D} \mathbf{W} \cdot d\mathbf{s}
\end{aligned}
$$

Hence,

$$
\int_{D} \mathbf{W} \cdot d\mathbf{s} = -\frac{2}{3}
$$

Problem 128

The surface S is a disk in the yz-plane. Its boundary is thus the unit circle in the yz-plane. The tricky part is determining the proper orientation on the boundary circle. A normal vector giving the orientation of S is determined by

$$
\frac{\partial \Psi}{\partial r} \times \frac{\partial \Psi}{\partial \theta} = \langle r, 0, 0 \rangle
$$

For $r \neq 0$ this is a vector which is parallel to the positive x-axis. The orientation on ∂S is then given by the right-hand rule, as in the following figure.

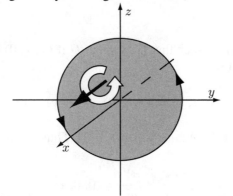

The usual parameterization for the unit circle in the yz-plane would be

$$\phi(t) = (0, \cos t, \sin t), \quad 0 \leq t \leq 2\pi$$

Notice that $\frac{d\phi}{dt} = (0, -\sin t, \cos t)$. At $t = 0$, $\phi(0) = (0, 1, 0)$, and $\frac{d\phi}{dt}(0) = (0, 0, 1)$. This agrees with the orientation given in the figure. We are now prepared to integrate, using Stokes' Theorem:

$$\int_S (\nabla \times \mathbf{W}) \cdot d\mathbf{S} = \int_{\partial S} \mathbf{W} \cdot d\mathbf{s}$$

$$= \int_0^{2\pi} (0, 0, \cos t) \cdot (0, -\sin t, \cos t) \; dt$$

$$= \int_0^{2\pi} \cos^2 t \; dt$$

$$= \left. \frac{1}{2}t + \frac{1}{4}\sin 2t \right|_0^{2\pi}$$

$$= \pi$$

As the chosen parameterization ϕ agrees with the orientation on ∂S there is no need to change sign.

Problem 129

Let Q be the region in $\mathbb{R}^2$ such that (x, y) is in Q if $(x, y, 0)$ is in S. Then we may parameterize S as follows:

$$\Psi(x, y) = (x, y, 0), \quad (x, y) \in Q$$

It follows that $\frac{\partial \Psi}{\partial x} = (1, 0, 0)$ and $\frac{\partial \Psi}{\partial y} = (0, 1, 0)$. Thus

$$\frac{\partial \Psi}{\partial x} \times \frac{\partial \Psi}{\partial y} = (0, 0, 1)$$

By definition

$$\int_S (\nabla \times \mathbf{W}) \cdot d\mathbf{S} = \int_Q \left\langle 0, 0, \frac{\partial g}{\partial x} - \frac{\partial f}{\partial y} \right\rangle \cdot \langle 0, 0, 1 \rangle \ dx \ dy$$

$$= \int_Q \frac{\partial g}{\partial x} - \frac{\partial f}{\partial y} \ dx \ dy$$

Also by definition

$$\int_{\partial S} \mathbf{W} \cdot d\mathbf{s} = \int_{\partial Q} \langle f(x, y), g(x, y) \rangle \cdot d\mathbf{s}$$

But Stokes' Theorem says

$$\int_S (\nabla \times \mathbf{W}) \cdot d\mathbf{S} = \int_{\partial S} \mathbf{W} \cdot d\mathbf{s}$$

So we get

$$\int_Q \frac{\partial g}{\partial x} - \frac{\partial f}{\partial y} \ dx \ dy = \int_{\partial Q} \langle f(x, y), g(x, y) \rangle \cdot d\mathbf{s}$$

which is Green's Theorem.

Problem 130

The boundary of S is parallel unit circles C_0 and C_1 in the planes $z = 0$ and $z = 1$, respectively. However, the orientations on these circles are opposite, as can be seen in the following figure.

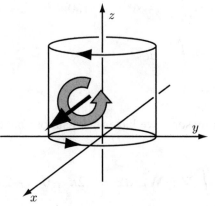

A parameterization for C_0 is given by

$$\Psi_0(t) = (\cos t, \sin t, 0), \quad 0 \leq t \leq 2\pi$$

At the point $\Psi_0(0) = (1, 0, 0)$ the derivative $\frac{d\Psi_0}{dt} = \langle -\sin t, \cos t, 0 \rangle$ is the vector $\langle 0, 1, 0 \rangle$. This agrees with the arrow in the picture.

Similarly, a parameterization for C_1 is given by

$$\Psi_1(t) = (\cos t, \sin t, 1), \quad 0 \leq t \leq 2\pi$$

At the point $\Psi_0(0) = (1, 0, 1)$ the derivative $\frac{d\Psi_0}{dt} = \langle -\sin t, \cos t, 0 \rangle$ is again the vector $\langle 0, 1, 0 \rangle$. This is opposite to the arrow in the picture, so we will have to remember to negate the integral over this curve.

Stokes' Theorem says

$$\int_S (\nabla \times \mathbf{W}) \cdot d\mathbf{S} = \int_{C_0 \cup C_1} \mathbf{W} \cdot d\mathbf{s}$$

$$= \int_{C_0} \mathbf{W} \cdot d\mathbf{s} + \int_{C_1} \mathbf{W} \cdot d\mathbf{s}$$

We do each of these integrals separately

$$\int_{C_0} \mathbf{W} \cdot d\mathbf{s} = \int_0^{2\pi} \langle 0, 0, 0 \rangle \cdot \langle -\sin t, \cos t, 0 \rangle \; dt = 0$$

$$\int_{C_1} \mathbf{W} \cdot d\mathbf{s} = \int_0^{2\pi} \langle -\sin t, \cos t, 0 \rangle \cdot \langle -\sin t, \cos t, 0 \rangle \; dt$$

$$= \int_0^{2\pi} dt$$

$$= 2\pi$$

But, for orientation reasons, we have to negate our answer, yielding -2π. Our final answer is thus

$$\int_S (\nabla \times \mathbf{W}) \cdot d\mathbf{S} = -2\pi + 0 = -2\pi$$

Problem 131

We assume the sphere is oriented with an outward normal (just reverse the picture if not). Dividing the sphere S into a northern hemisphere S_+ and a southern hemisphere S_-, we see in the following figure that the equator gets opposite orientation as the boundary of each.

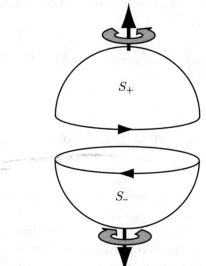

As the induced orientations on the equator are opposite, it follows that

$$\int_{\partial S_+} \mathbf{W} \cdot d\mathbf{s} = - \int_{\partial S_-} \mathbf{W} \cdot d\mathbf{s}$$

We now employ Stokes' Theorem

$$\int_S (\nabla \times \mathbf{W}) \cdot d\mathbf{S} = \int_{S_+} (\nabla \times \mathbf{W}) \cdot d\mathbf{S} + \int_{S_-} (\nabla \times \mathbf{W}) \cdot d\mathbf{S}$$

$$= \int_{\partial S_+} \mathbf{W} \cdot d\mathbf{s} + \int_{\partial S_-} \mathbf{W} \cdot d\mathbf{s}$$

$$= 0$$

Problem 132

The direction of $\nabla \times \mathbf{W}$ will be perpendicular to the plane containing the loop with greatest circulation. Since this is the xz-plane, the direction of $\nabla \times \mathbf{W}$ will be parallel to the y-axis. The magnitude of $\nabla \times \mathbf{W}$ can be estimated by

$$|\nabla \times \mathbf{W}(p)| \approx \frac{1}{\text{Area}(D)} \int_C \mathbf{W} \cdot d\mathbf{s}$$

where D is the disk whose boundary is the loop C. Since the radius of C is .1, the area of D is $\pi(.1)^2$. This gives us

$$\frac{1}{\text{Area}(D)} \int_C \mathbf{W} \cdot d\mathbf{s} = \frac{1}{\pi(.1)^2}(.5) \approx 15.92$$

We conclude that at the origin

$$\nabla \times \mathbf{W} \approx \langle 0, 15.92, 0 \rangle$$

Problem 133

The circulation of the vector field around any loop in a plane parallel to the xz-plane would be zero, since the vector field is constant on such a plane. A vector V that is perpendicular to such a loop points in the y-direction. The circulation around any horizontal loop would also have to be zero, since the vector field is perpendicular to such a loop. A vector W that is perpendicular to such a loop points in the z-direction. The curl must then be perpendicular to both V and W, and so must point in the x-direction.

One can also see this algebraically. Such a vector field must look like $\langle 0, 0, f(y) \rangle$. The curl of this is

$$\begin{vmatrix} \mathbf{i} & \mathbf{j} & \mathbf{k} \\ \frac{\partial}{\partial x} & \frac{\partial}{\partial y} & \frac{\partial}{\partial z} \\ 0 & 0 & f(y) \end{vmatrix} = \langle f'(y), 0, 0 \rangle$$

Problem 134

Let V be the unit cube. Then Gauss' Theorem tells us

$$\int_{\partial V} \mathbf{W} \cdot d\mathbf{S} = \int \int_V \int \nabla \cdot \mathbf{W} \, dx \, dy \, dz$$

$$= \int_0^1 \int_0^1 \int_0^1 \nabla \cdot \mathbf{W} \, dx \, dy \, dz$$

$$= \int_0^1 \int_0^1 \int_0^1 2xyz + 2xyz + 2xyz \, dx \, dy \, dz$$

$$= \int_0^1 \int_0^1 \int_0^1 6xyz \, dx \, dy \, dz$$

$$= \int_0^1 \int_0^1 3yz \, dy \, dz$$

$$= \int_0^1 \frac{3}{2} z \, dz$$

$$= \frac{3}{4}$$

Problem 135

By Gauss' Theorem the integral of $\nabla \cdot \mathbf{W}$ over the ball B is equal to the integral of $\mathbf{W}$ over the unit sphere S, with outward-pointing normal vector. The unit sphere is parameterized in the usual way by

$$\Psi(\theta, \phi) = (\sin \phi \cos \theta, \sin \phi \sin \theta, \cos \phi)$$

$$0 \le \theta \le 2\pi, \quad 0 \le \phi \le \pi$$

As in Example 11-8,

$$\frac{\partial \Psi}{\partial \theta} \times \frac{\partial \Psi}{\partial \phi} = (-\sin^2 \phi \cos \theta, -\sin^2 \phi \sin \theta, -\sin \phi \cos \phi)$$

which agrees with the orientation on S.

We now compute

$$\iiint_B \nabla \cdot \mathbf{W} \, dx \, dy \, dz = \int_S \mathbf{W} \cdot d\mathbf{S}$$

$$= \int_0^\pi \int_0^{2\pi} \langle 0, 0, e^{\cos \phi} \rangle \cdot \langle -\sin^2 \phi \cos \theta, -\sin^2 \phi \sin \theta,$$

$$- \sin \phi \cos \phi \rangle \, d\theta \, d\phi$$

$$= \int_0^\pi \int_0^{2\pi} -\sin\phi \cos\phi\, e^{\cos\phi}\, d\theta\, d\phi$$

$$= 2\pi \int_0^\pi -\sin\phi \cos\phi\, e^{\cos\phi}\, d\phi$$

$$= 2\pi \int_1^{-1} u e^u\, du$$

$$= 2\pi (u e^u - e^u)|_1^{-1}$$

$$= -\frac{4\pi}{e}$$

Problem 136

The region V is parameterized by

$$\Psi(r, \theta, z) = (r\cos\theta, r\sin\theta, z)$$

$$1 \le r \le 2, \quad 0 \le \theta \le \frac{\pi}{2}, \quad 0 \le z \le 2$$

We use Gauss' Theorem to transform the integral:

$$\int_{\partial V} \mathbf{W} \cdot d\mathbf{S} = \int \int_V \int \nabla \cdot \mathbf{W}\, dx\, dy\, dz$$

$$= \int \int_V \int 3(x^2 + y^2)\, dx\, dy\, dz$$

We now use the parameterization to change variables:

$$\int \int_V \int 3(x^2 + y^2)\, dx\, dy\, dz$$

$$= \int_0^2 \int_0^{\frac{\pi}{2}} \int_1^2 3((r\cos\theta)^2 + (r\sin\theta)^2) \begin{vmatrix} \cos\theta & \sin\theta & 0 \\ -r\sin\theta & r\cos\theta & 0 \\ 0 & 0 & 1 \end{vmatrix} dr\, d\theta\, dz$$

$$= \int_0^2 \int_0^{\frac{\pi}{2}} \int_1^2 3r^2(r) \, dr \, d\theta \, dz$$

$$= \int_0^2 \int_0^{\frac{\pi}{2}} \int_1^2 3r^3 \, dr \, d\theta \, dz$$

$$= \int_0^2 \int_0^{\frac{\pi}{2}} \frac{3}{4} r^4 \Big|_1^2 \, d\theta \, dz$$

$$= \int_0^2 \int_0^{\frac{\pi}{2}} \frac{45}{4} \, d\theta \, dz$$

$$= \frac{45\pi}{4}$$

Problem 137

First, we observe that the *polar* equation $r = \cos\theta$ is a circle of radius $\frac{1}{2}$, centered on the point $(\frac{1}{2}, 0)$. Hence, the surface C is a cylinder of height 2. The surface D is a disk, which caps off the top of the cylinder C, like an upside-down can. Let E denote the "bottom" of the can. That is, E is the set of points in the xy-plane which are within $\frac{1}{2}$ of a unit away from the point $(\frac{1}{2}, 0, 0)$. We assume an orientation is given on E so that C, D, and E together form the (oriented) boundary of V, the points inside the "can." Applying Gauss' Theorem gives us

$$\iiint_V \nabla \cdot \mathbf{W} \, dx \, dy \, dz = \int_{C+D+E} \mathbf{W} \cdot d\mathbf{S}$$

$$= \int_{C+D} \mathbf{W} \cdot d\mathbf{S} + \int_E \mathbf{W} \cdot d\mathbf{S}$$

Now notice that on E the vector field $\mathbf{W} = \langle 0, 0, 0 \rangle$, so

$$\int_E \mathbf{W} \cdot d\mathbf{S} = 0$$

Putting these results together gives us

$$\iiint_V \nabla \cdot \mathbf{W} \, dx \, dy \, dz = \int_{C+D} \mathbf{W} \cdot d\mathbf{S}$$

We may thus obtain the desired answer by evaluating the integral on the left-hand side of this last equation. The first thing we will need to evaluate this is the divergence of $\mathbf{W}$:

$$\nabla \cdot \mathbf{W} = \frac{\partial}{\partial y} xyz = xz$$

Next, we will need to parameterize V. Notice that V is a solid cylinder of radius $\frac{1}{2}$, whose central axis has been translated $\frac{1}{2}$ of a unit away from the z-axis, in the positive x-direction. A parameterization is thus given by

$$\Psi(r, \theta, z) = \left(r \cos \theta + \frac{1}{2}, r \sin \theta, z \right)$$

$$0 \le r \le \frac{1}{2}, \quad 0 \le \theta \le 2\pi, \quad 0 \le z \le 1$$

The partials of this parameterization are

$$\frac{\partial \Psi}{\partial r} = \langle \cos \theta, \sin \theta, 0 \rangle$$

$$\frac{\partial \Psi}{\partial \theta} = \langle -r \sin \theta, r \cos \theta, 0 \rangle$$

$$\frac{\partial \Psi}{\partial z} = \langle 0, 0, 1 \rangle$$

The determinant of the matrix which consists of these vectors is

$$\begin{vmatrix} \cos \theta & \sin \theta & 0 \\ -r \sin \theta & r \cos \theta & 0 \\ 0 & 0 & 1 \end{vmatrix} = r$$

We now use Ψ to evaluate the integral:

$$\iiint_V \nabla \cdot \mathbf{W}\, dx\, dy\, dz \quad \iiint_V xz\, dx\, dy\, dz$$

$$= \int_0^1 \int_0^{2\pi} \int_0^{\frac{1}{2}} \left(r \cos\theta + \frac{1}{2} \right)(z)(r)\, dr\, d\theta\, dz$$

$$= \int_0^1 \int_0^{2\pi} \int_0^{\frac{1}{2}} r^2 z \cos\theta + \frac{rz}{2}\, dr\, d\theta\, dz$$

$$= \int_0^1 \int_0^{2\pi} \frac{1}{3}\left(\frac{1}{2}\right)^3 z \cos\theta + \frac{1}{4}\left(\frac{1}{2}\right)^2 z\, d\theta\, dz$$

$$= \int_0^1 \int_0^{2\pi} \frac{1}{24} z \cos\theta + \frac{1}{16} z\, d\theta\, dz$$

$$= \int_0^1 \frac{\pi}{8} z\, dz$$

$$= \frac{\pi}{16}$$

Problem 138

S_1, and S_2 *with the opposite orientation*, together bound a volume V of $\mathbb{R}^3$.

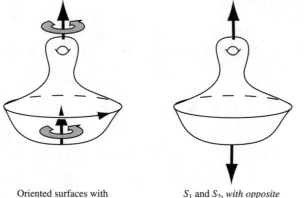

Oriented surfaces with
the same oriented boundary

S_1 and S_2, *with opposite*
orientation, bound a volume V

By Gauss' Theorem

$$\iiint_V \nabla \cdot \mathbf{W}\, dx\, dy\, dz = \int_{\partial V} \mathbf{W} \cdot d\mathbf{S}$$

$$= \int_{S_1 - S_2} \mathbf{W} \cdot d\mathbf{S}$$

$$= \int_{S_1} \mathbf{W} \cdot d\mathbf{S} - \int_{S_2} \mathbf{W} \cdot d\mathbf{S}$$

But $\nabla \cdot \mathbf{W} = 0$ implies $\iiint_V \nabla \cdot \mathbf{W}\, dx\, dy\, dz = 0$. We conclude

$$\int_{S_1} \mathbf{W} \cdot d\mathbf{S} = \int_{S_2} \mathbf{W} \cdot d\mathbf{S}$$

Problem 139

$$\nabla \cdot \mathbf{W}(p) \approx \frac{1}{\text{Volume}(B)} \int_{\partial B} \mathbf{W} \cdot d\mathbf{S}$$

$$= \frac{1}{\frac{4}{3}\pi (.1)^3}(.5)$$

$$\approx 119.36662$$

Chapter 11 Quiz

Problem 140

1. The key to this problem is to notice that $\mathbf{W} = \nabla f(x, y, z)$, where $f(x, y, z) = xy^2z^2$. So,

$$\int_C \mathbf{W} \cdot d\mathbf{s} = \int_C (\nabla f) \cdot d\mathbf{s}$$

$$= f(1, 1, 1) - f(0, 0, 0)$$

$$= 1 - 0$$

$$= 1$$

2. Using Green's Theorem

$$\int_{\partial \sigma} \langle 1, \ln x \rangle \cdot d\mathbf{s} = \int \int_{\sigma} \frac{\partial}{\partial x}(-\ln x) - \frac{\partial}{\partial y}(1) \, dx \, dy$$

$$= \int \int_{\sigma} -\frac{1}{x} \, dx \, dy$$

To evaluate this integral we will need the partials of the parameterization:

$$\frac{\partial \phi}{\partial u} = \langle v^2, 3u^2 v \rangle$$

$$\frac{\partial \phi}{\partial v} = \langle 2uv, u^3 \rangle$$

The determinant of the matrix of partials is thus

$$\begin{vmatrix} v^2 & 3u^2 v \\ 2uv & u^3 \end{vmatrix} = u^3 v^2 - 6u^3 v^2 = -5u^3 v^2$$

We now integrate:

$$\int \int_{\sigma} -\frac{1}{x} \, dx \, dy = \int_1^2 \int_1^2 \frac{-1}{uv^2}(-5u^3 v^2) \, du \, dv$$

$$= \int_1^2 \int_1^2 5u^2 \, du \, dv$$

$$= \int_1^2 \frac{5}{3} u^3 \Big|_1^2 \, dv$$

$$= \int_1^2 \frac{35}{3} \, dv$$

$$= \frac{35}{3}$$

3. The curves C_1 and C_2 lie in the xy-plane, and are pictured below.

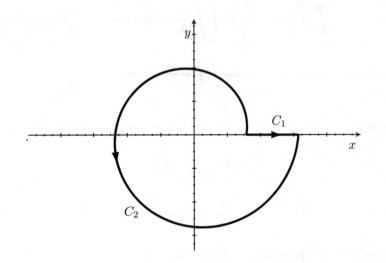

Let C_1^- denote the curve C_1 with the opposite orientation. Then the curves C_1^- and C_2 together form the (oriented) boundary of a region R of the xy-plane. Stokes' Theorem then says

$$\int_{C_1^- \cup C_2} \mathbf{W} \cdot d\mathbf{s} = \int_R (\nabla \times \mathbf{W}) \cdot d\mathbf{S}$$

But the statement of the problem specifies $\nabla \times \mathbf{W} = \langle 0, 0, 0 \rangle$, so the integral on the right (and hence the integral on the left) is 0. Now notice

$$\int_{C_1^- \cup C_2} \mathbf{W} \cdot d\mathbf{s} = \int_{C_1^-} \mathbf{W} \cdot d\mathbf{s} + \int_{C_2} \mathbf{W} \cdot d\mathbf{s}$$

$$= \int_{C_2} \mathbf{W} \cdot d\mathbf{s} - \int_{C_1} \mathbf{W} \cdot d\mathbf{s}$$

Since this is zero, the desired result follows.

Final Exam

Problem 141

1.(a)

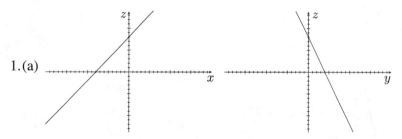

(b) To find the critical points we set the partial derivatives equal to zero:

$$\frac{\partial f}{\partial x} = y + 1 = 0$$

$$\frac{\partial f}{\partial y} = x - 2 = 0$$

The first equation tells us $y = -1$ and the second tells us $x = 2$. So $(2, -1)$ is the only critical point. Now we compute

$$\begin{vmatrix} \dfrac{\partial^2 f}{\partial x^2} & \dfrac{\partial^2 f}{\partial x \, \partial y} \\ \dfrac{\partial^2 f}{\partial y \, \partial x} & \dfrac{\partial^2 f}{\partial y^2} \end{vmatrix} = \begin{vmatrix} 0 & 1 \\ 1 & 0 \end{vmatrix} = -1$$

Since this is negative the critical point $(2, -1)$ corresponds to a saddle.

(c) First, we find a unit vector that points in the desired direction:

$$\frac{\langle 1, 2 \rangle}{|\langle 1, 2 \rangle|} = \frac{\langle 1, 2 \rangle}{\sqrt{5}} = \left\langle \frac{\sqrt{5}}{5}, \frac{2\sqrt{5}}{5} \right\rangle$$

The desired slope is the directional derivative in this direction:

$$\nabla_{\left(\frac{\sqrt{5}}{5}, \frac{2\sqrt{5}}{5} \right)} f(0, 1) = \left\langle \frac{\sqrt{5}}{5}, \frac{2\sqrt{5}}{5} \right\rangle \cdot \nabla f(0, 1)$$

$$= \left\langle \frac{\sqrt{5}}{5}, \frac{2\sqrt{5}}{5} \right\rangle \cdot \langle 2, -2 \rangle$$

$$= \frac{-2\sqrt{5}}{5}$$

(d)   Volume $= \displaystyle\int_0^2 \int_0^1 xy + x - 2y + 4 \, dx \, dy$

$$= \int_0^2 \frac{1}{2}x^2 y + \frac{1}{2}x^2 - 2yx + 4x \Big|_0^1 \, dy$$

$$= \int_0^2 \frac{1}{2}y + \frac{1}{2} - 2y + 4 \, dy$$

$$= \frac{1}{4}y^2 + \frac{1}{2}y - y^2 + 4y \Big|_0^2$$

$$= 1 + 1 - 4 + 8$$

$$= 6$$

2. Since $r^2 = x^2 + y^2$, the desired surface has the cylindrical equation $z = 4 - r^2$. Where the graph hits the xy-plane we know $z = 0$, and hence $0 = 4 - r^2$, or $r = 2$. A parameterization is thus given utilizing cylindrical coordinates by

$$\Psi(r, \theta) = (r \cos \theta, r \sin \theta, 4 - r^2)$$

$$0 \le r \le 2, \quad 0 \le \theta \le 2\pi$$

3. There are several ways to do this. We will utilize spherical coordinates, since that is how the problem was stated. Being on a sphere of radius 1 says $\rho = 1$. We now plug this, and the information $\theta = \phi$, into the usual spherical coordinates:

$$\Psi(\theta) = (\sin \theta \cos \theta, \sin \theta \sin \theta, \cos \theta)$$

To get the whole circle the domain should be $0 \le \theta \le 2\pi$.

4. $\nabla \cdot \mathbf{W} = \dfrac{\partial}{\partial x} xz^2 + \dfrac{\partial}{\partial z} xz^2$

$$= z^2 + 2xz$$

$$\nabla \times \mathbf{W} = \begin{vmatrix} \mathbf{i} & \mathbf{j} & \mathbf{k} \\ \frac{\partial}{\partial x} & \frac{\partial}{\partial y} & \frac{\partial}{\partial z} \\ xz^2 & 0 & xz^2 \end{vmatrix}$$

$$= \langle 0, 2xz - z^2, 0 \rangle$$

5. Since the region V is cylindrical, this is best done by utilizing a parameterization. We parameterize V in the usual way with cylindrical coordinates:

$$\Psi(r, \theta, z) = (r \cos \theta, r \sin \theta, z)$$

$$0 \leq r \leq 1, \quad 0 \leq \theta \leq \frac{\pi}{2}, \quad 0 \leq z \leq 1$$

To do the integral we will need the determinant of the matrix of partial derivatives, which simplifies to r. We can thus integrate as follows:

$$\int_V 2\sqrt{1 + x^2 + y^2}\, dx\, dy\, dz = \int_0^1 \int_0^{\frac{\pi}{2}} \int_0^1 2 \left(\sqrt{1 + (r \cos \theta)^2 + (r \sin \theta)^2} \right) (r)\, dr\, d\theta\, dz$$

$$= \int_0^1 \int_0^{\frac{\pi}{2}} \int_0^1 2r\sqrt{1 + r^2}\, dr\, d\theta\, dz$$

$$= \int_0^1 \int_0^{\frac{\pi}{2}} \int_1^2 \sqrt{u}\, du\, d\theta\, dz$$

$$= \int_0^1 \int_0^{\frac{\pi}{2}} \frac{2}{3}(2^{\frac{3}{2}} - 1)\, d\theta\, dz$$

$$= \int_0^1 \frac{\pi}{3}(2^{\frac{3}{2}} - 1)\, dz$$

$$= \frac{\pi}{3}(2^{\frac{3}{2}} - 1)$$

6. The derivative of the parameterization is

$$\frac{d\phi}{dt} = \langle -2 \sin t, 2 \cos t, 2t \rangle$$

We now integrate:

$$\int_C \langle 0, 0, x^2 + y^2 \rangle \cdot d\mathbf{s} = \int_0^2 \langle 0, 0, (2\cos t)^2 + (2\sin t)^2 \rangle \cdot \langle -2\sin t, 2\cos t, 2t \rangle \; dt$$

$$= \int_0^2 \langle 0, 0, 4 \rangle \cdot \langle -2\sin t, 2\cos t, 2t \rangle \; dt$$

$$= \int_0^2 8t \; dt$$

$$= 4t^2 \Big|_0^2$$

$$= 16$$

7. First, we compute the partials of the parameterization:

$$\frac{\partial \phi}{\partial r} = \langle \cos\theta, \sin\theta, 2r \rangle$$

$$\frac{\partial \phi}{\partial \theta} = \langle -r\sin\theta, r\cos\theta, 0 \rangle$$

The cross product of these vectors is

$$\frac{\partial \phi}{\partial r} \times \frac{\partial \phi}{\partial \theta} = \begin{vmatrix} \mathbf{i} & \mathbf{j} & \mathbf{k} \\ \cos\theta & \sin\theta & 2r \\ -r\sin\theta & r\cos\theta & 0 \end{vmatrix} = \langle -2r^2\cos\theta, -2r^2\sin\theta, r \rangle$$

We now integrate:

$$\int_P \mathbf{F} \cdot d\mathbf{S} = \int_0^{\frac{\pi}{2}} \int_0^1 \langle 0, -r^2, 0 \rangle \cdot \langle -2r^2\cos\theta, -2r^2\sin\theta, r \rangle \; dr \; d\theta$$

$$= \int_0^{\frac{\pi}{2}} \int_0^1 2r^4 \sin\theta \; dr \; d\theta$$

$$= \int_0^{\frac{\pi}{2}} \frac{2}{5} \sin\theta \; d\theta$$

$$= \frac{2}{5}$$

8. The key to this problem is to notice that $\mathbf{W} = \nabla f$, where $f(x, y, z) = xyz$. Then, using the independence of path of line integrals

$$\int_C \mathbf{W} \cdot d\mathbf{s} = \int_C (\nabla f) \cdot d\mathbf{s}$$

$$= f\left(\phi\left(\frac{\pi}{4}\right)\right) - f(\phi(0))$$

$$= f\left(\frac{\sqrt{2}}{2}, \frac{\sqrt{2}}{2}, 1\right) - f(0, 0, 0)$$

$$= \frac{1}{2}$$

9. The surface S bounds a volume V which is parameterized by

$$\Psi(r, \theta, z) = (r \cos \theta, r \sin \theta, z)$$

$$0 \leq r \leq 1, \quad 0 \leq \theta \leq 2\pi, \quad 0 \leq z \leq 1$$

To evaluate an integral over this region we will need the determinant of the matrix of partials. We have done this calculation several times. The reader may check that the answer is r. We now integrate

$$\int_S \mathbf{W} \cdot d\mathbf{S} = \int \int_V \int \nabla \cdot \mathbf{W} \, dx \, dy \, dz$$

$$= \int \int_V \int 2z \, dx \, dy \, dz$$

$$= \int_0^1 \int_0^{2\pi} \int_0^1 2z(r) \, dr \, d\theta \, dz$$

$$= \int_0^1 \int_0^{2\pi} z \, d\theta \, dz$$

$$= \int_0^1 2\pi z \, dz$$

$$= \pi$$

INDEX